Machine Component Design

Machine Component Design

Editor
Bianca Lupei

Machine Component Design
Edited by **Bianca Lupei**

Printed in 2017

ISBN: 978-1-68117-161-6
Library of Congress Control Number: 2015950370

© 2016 by
SCITUS Academics LLC,
616, Corporate Way, Suite 2, 4766,
Valley Cottage, NY 10989

www.scitusacademics.com

This book contains information obtained from highly regarded resources. Copyright for individual articles remains with the authors as indicated. All chapters are distributed under the terms of the Creative Commons Attribution License, which permits unrestricted use, distribution, and reproduction in any medium, provided the original author and source are credited.

Notice

Reasonable efforts have been made to publish reliable data and views articulated in the chapters are those of the individual contributors, and not necessarily those of the editors or publishers. Editors or publishers are not responsible for the accuracy of the information in the published chapters or consequences of their use. The publisher believes no responsibility for any damage or grievance to the persons or property arising out of the use of any materials, instructions, methods or thoughts in the book. The editors and the publisher have attempted to trace the copyright holders of all material reproduced in this publication and apologize to copyright holders if permission has not been obtained. If any copyright holder has not been acknowledged, please write to us so we may rectify.

Table of Content

CHAPTER	1	Survey of Mechatronic Techniques in Modern Machine Design	1
CHAPTER	2	Method for Remanufacturing Large-Sized Skew Bevel Gears Using CNC Machining Center	21
CHAPTER	3	Circular Causality and Indeterminism in Machines for Design	55
CHAPTER	4	Micro- and Nano-Air Vehicles: State of the Art	73
CHAPTER	5	Radial Ball Bearings with Angular Contact in Machine Tools	117
CHAPTER	6	Cryogenic Tribology in High–Speed Bearings and Shaft Seals of Rocket Turbo Pumps	169
CHAPTER	7	Surface Modification by Friction Based Processes	223
CHAPTER	8	Micro/Nano-Mechanical Sensors and Actuators Based on Soi-Mems Technology	247
		Index	265

Preface

A machine has a power source and actuators that generate forces and movement, and a system of mechanisms that shape the actuator input to achieve a specific application of output forces and movement. Machine component refers to an elementary component of a machine.

Machine component may be features of a part (such as screw threads or integral plain bearings) or they may be discrete parts in and of themselves such as wheels, axles, pulleys, rolling-element bearings, or gears. All of the simple machines may be described as machine elements, and many machine elements incorporate concepts of one or more simple machines.

The book, Machine Component Design, involves analytical methodologies for determining strength, stiffness and stability of a mechanical component and application of these methodologies to determine the size, shape, geometry and life of the components. Intended to serve as a reference tool on design of machine elements for students in mechanical, production and industrial engineering as well as for practicing engineers, this book is focused on all aspects of design of machine components including material selection and lift or performance estimation under static. Fatigue, impact and creep loading conditions. The wide range of real life applications and examples presented in the book provide conceptual understanding of complex and important engineering theories and will help students and practitioners to improve the decision process in the field of mechanical component design.

CHAPTER 1

Survey of Mechatronic Techniques in Modern Machine Design

Devdas Shetty[1], Lou Manzione[2] and Ahad Ali[3]

[1]School of Engineering and Applied Sciences, University of the District of Columbia, Washington, DC 20008, USA
[2]Department of Mechanical Engineering, University of Hartford, West Hartford, CT 06117, USA
[3]A. Leon Linton Department of Mechanical Engineering, Lawrence Technological University, Southfield, MI 48188, USA

ABSTRACT

Increasing demands on the productivity of complex systems, such as manufacturing machines and their steadily growing technological importance will require the application of new methods in the product development process. A smart machine can make decisions about the process in real-time with plenty of adaptive controls. This paper shows the simulation based mechatronic model of a complex system with a better understanding of the dynamic behavior and interactions of the components. This offers improved possibilities of evaluating and optimizing the dynamic motion performance of the entire automated system in the early stages of the design process. Another effect is the growing influence of interactions between machine components on achievable machine dynamics and precision and quality of components. The examples cited in this paper, demonstrate the distinguishing feature of mechatronic systems through intensive integration. The case studies also show that it will no longer be sufficient to focus solely on the optimization of subsystems. Instead it will be necessary to strive for optimization of the complete system. The interactions between machine components, the influence of the control system and the machining process will have to be considered during the design process and the coordination of feed drives and frame structure components.

SIMULATION OF COMPLEX STRUCTURES

Mechatronics is a methodology used for the optimal design of electromechanical products. The term was coined nearly 40 years ago, in 1969, when the engineer Tetsuro Mori combined the words "mechanical" and "electronic" to describe the electronic control systems that Yaskawa Electric Corporation was building for the mechanical factory equipment. Mechatronics is a design philosophy, which is an integrating approach to an engineering design as shown in Figure 1. The primary factor in mechatronics is the involvement of these areas throughout the design process. Through a mechanism of simulating interdisciplinary ideas and techniques, mechatronics provides ideal conditions to raise the synergy, thereby providing a catalytic effect for the new solutions to technically complex situations. An important characteristic of the mechatronic devices and systems is their built-in intelligence that results through a combination of precision mechanical and electrical engineering and real-time programming integrated to the design process. The synergy can be generated by the right combination of parameters; that is, the final product can be better than just the sum of its parts. Mechatronic products exhibit performance characteristics that were previously difficult to be achieved without the synergistic combination. Recently, some mechatronic applications are presented on micro motion and helicopter and robotic arm in different research articles [1–5].

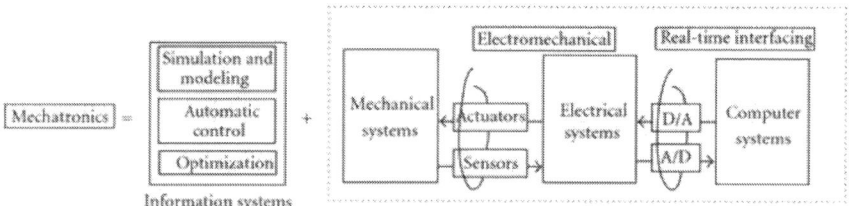

Figure 1. Mechatronic Process.

A typical mechatronic design process [6] is shown in Figure 2. Starting with steps 5 through 9, software tools are available to aid the designer in creating and debugging the mathematical system models. Some tools that are particularly useful allow the designer to represent the system by creating a system block diagram from simple building blocks such as integrators, gain stages, summing junctions, and nonlinear switches. Some examples of these tools are LabVIEW, Simulink/Matlab, Matrix-x, ACSL, SimPACK, Hypersignal, and VisSim. These graphical simulation tools run on generic platforms such as desktop PC compatible Windows operating systems. With any of these tools, the designer can create a plant model and then validate it against real-world measurements (step 5). Once the plant model has been validated, the designer can then design the control system and optimize it until the correct response is achieved (steps 6 and 7). In some cases, completely accurate plant models cannot be made and certain assumptions must be made about that plant model that cannot be validated. In these cases, it is advantageous to be able to test the control system within the plant environment (step 8). This is sometimes referred to as hardware-in-the-loop simulation since some of the actual hardware (mechanical and electrical parts) is used in the system control loop (acting as the plant that is to be controlled). The hardware-in-the-loop simulation testing provides the designer with reassurance that any assumptions made on the plant model were correct. If any assumptions were incorrect, however, the designer has the opportunity to optimize the design (step 9) before committing.

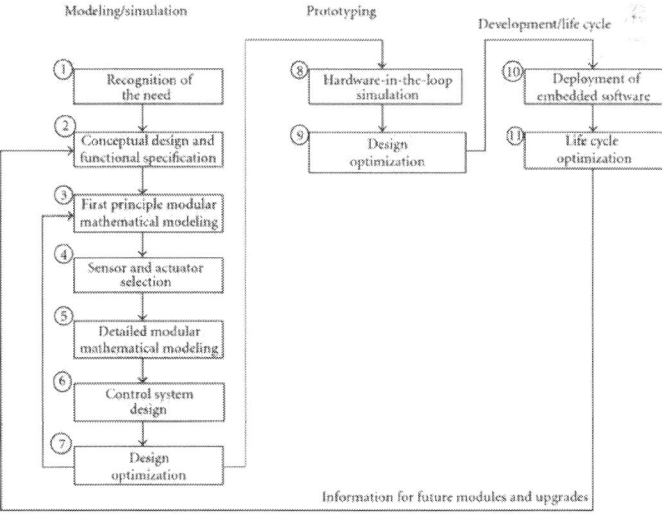

Figure 2. Mechatronic design [6].

DESIGN OF SMART MACHINE TOOLS USING MECHATRONICS

Researchers used mechatronic techniques for different applications. One of the applications is for smart machine tool design [7]. It explains the application of integrated CAx tools for setting up a virtual prototype that will permit evaluation and optimization of the entire machine tool's motion dynamics in early phases of the development process. Due to the development of machine design and drive technology, modern numerically controlled machine tools can be described to an increasing extent as characteristic examples of complex mechatronic systems [8]. A distinguishing feature of mechatronic systems is the achievement of system functionality through intensive integration of electronical and information sub functions on a mechanical carrier [9].

In another research, tool path interpolation of the NC motion control is of great importance for the obtainable motion dynamics and the resulting contouring error, especially for highly dynamic machine movements. The contouring error consists of the tracking error of the feed axes and of the deflections of the TCP caused by the physical effects of the flexible machine structure. In order to reduce the contouring error, modern NC controllers make use of two major technological approaches [10].

Optical technology has been incorporated into mechatronic systems at an accelerated rate, and as a result, a great number of machines/systems with smart optical components have been presented which introduces the fundamental concept, definition, and characteristics of the technology by analyzing the characteristics of a variety of practical optomechatronic systems [11]. Optical elements have been increasingly incorporated at an accelerated rate into mechatronic systems, and vice versa [12–19].

Optomechatronics has its roots in technological developments of mechatronics and optoelectronics. Figure 3shows the chronology of those developments [20]. In the 1960s, the electronic revolution came with the integration of transistor and other semiconductor devices into monolithic circuits, and, in 1971, the semiconductor fabrication technology brought about a tremendous impact on a broad spectrum of technological fields. In the 1980s, the semiconductor technology also created microelectromechanical systems (MEMS), and this brought about a new dimension of machines/systems, micro sizing their dimension.

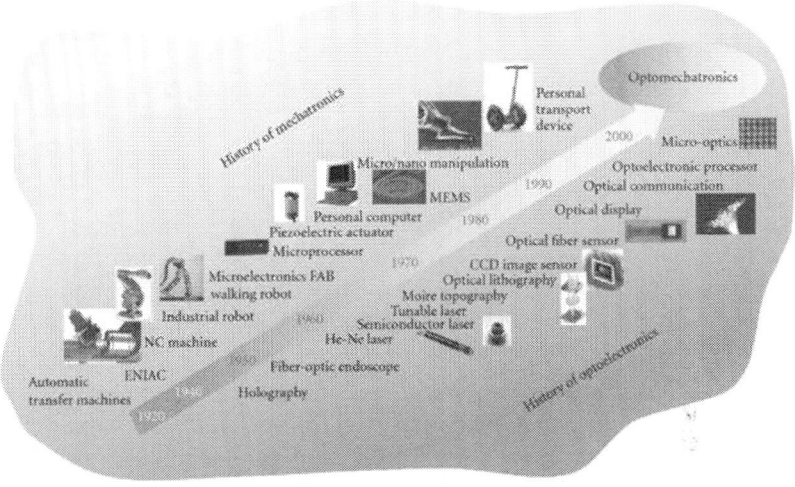

Figure 3. History of optomechatronics [20].

The major functions and roles of mechatronic elements in optomechatronic systems can be categorized into the following five technological domains [21]: sensing, actuation, information feedback, motion/state control, and embedded intelligence with microprocessor.

In the last few years the Virtual Design of machine tools has been extensively studied in several manufacturing automation and production engineering laboratories of universities and research institutes. This new technology is mainly studied and applied to machining centers (MCs) for high speed milling (HSM), the manufacturing of complex dies, moulds and aerospace parts currently being a strategic sector of production engineering. Almost a hundred relevant papers [7, 22–25] have recently been published on these topics in leading engineering journals and presented at several technical conferences; this confirms the significant interest, both industrial and academic, in virtual design.

For CNC manufacturers and MC users, however, the full exploitation of virtual machine tool technology still requires [22] fundamental developments mainly in the areas of cutting process simulation and full integration of all analysis modules in a user-friendly environment. The integration of the two models in one simulation environment is now possible, making it possible to study the interactions between the dynamics of these active and passive mechanical structures [22, 26]. The optimization of their performances is a basic prerequisite for ensuring productivity at shop floor level: fast machining times, the required dimensional accuracy, and good surface quality of the work pieces. Gurbuz [27] presented the mechatronic approach for desktop CNC milling machine design.

Siemens introduced mechatronics concept designer with a new integrated machine design solution that develops and markets machine tools and production machines [28]. The mechatronics concept designer represents a paradigm shift for the industry with a new system engineering approach to machine design that captures "voice of the customer" input, manages early requirements, and facilitates the simultaneous definition and simulation of the complex mechanical, electrical, and automation software found in today's increasingly complex machine tools. With an easy-to-use, interactive simulation capability based on groundbreaking "gaming" technology, the mechatronics concept designer can help significantly reduce development time and improve product quality for the global machine design industry.

INTEGRATED DESIGN ISSUES IN MECHATRONICS

The integration within a mechatronic system can be performed through the combination of hardware (components) and software (information processing). Hardware integration results from designing the mechatronic system as an overall system and bringing together the sensors, actuators, and microcomputers into the mechanical system. Software integration is primarily based on advanced control functions. Figure 4illustrates how the hardware and software integration takes place. It also shows how an additional contribution of the process knowledge and information processing is involved beside the feedback process. A mechatronic product can achieve impressive results if it is effectively integrated with the concurrent engineering management strategy. The benefits that accrue are greater productivity, higher quality, production reliability by the incorporation of intelligent, and self-correcting sensory and feedback system. The basic approach involves accurate computer-based dynamic models from illustrations and other information using the analogy approach. This unique method combines the standard analogy approach to modeling, with block diagrams, the major difference being the ability to incorporate nonlinearities directly into the system without linearization.

Control system design methods are available with several design procedures for common control structures including PID, Lag, Lead, Rate Feedback, and pure gain. Signal processing and data interpretation are also handled using the visual programming approach. The hardware-in-the-loop simulation testing provides the designer with reassurance that any assumptions made on the plant model were correct. If any assumptions were incorrect, however, the designer does have the opportunity to optimize the design before committing to the real target hardware platform.

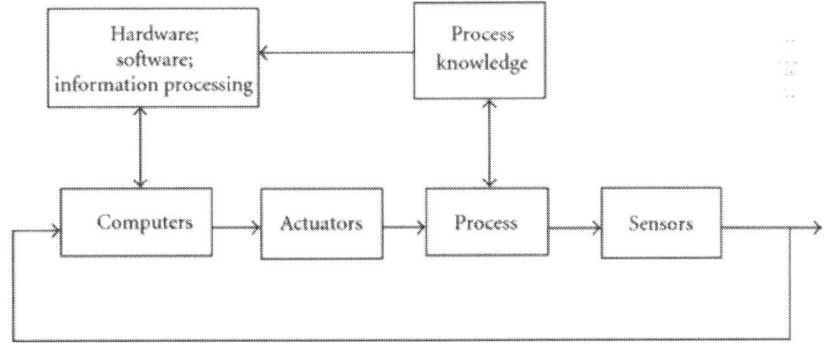

Figure 4. Hardware and software integration.

There are two main methods currently used to accomplish the hardware-in-the-loop simulation testing. One utilizes the desktop/workstation graphical user interface (GUI) coupled with standard data acquisition and control (DAC) interface circuit boards. One major drawback that simulation software and other PC based simulation systems suffer from is the inability to work in systems where loop responses need to be fast. Therefore once the control algorithms are designed and debugged, the algorithms must then be reimplemented, retested, and debugged on an embedded platform. The second method for hardware-in-the-loop testing involves cross-compiling the control algorithm to target an embedded real-time processor platform. The embedded processor platform is often a digital signal processor (DSP), with I/O that is tailored for embedded system products. Embedded processor platforms are designed for reduced cost and increased speed, and as such they generally do not have video displays nor standard desktop inputs such as full function keyboards and mouse interfaces. Figure 5 shows a typical setup for a DSP-based hardware-in-the-loop testing.

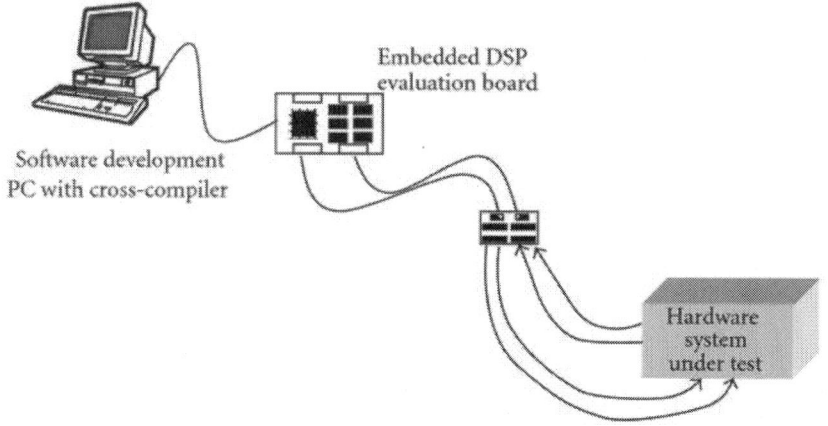

Figure 5. Embedded DSP-based "hardware-in-the-loop simulation."

The main elements of mechatronic approach are as follows:

a) Demonstration of a language-neutral approach to code system development. This is done using visual simulation.
b) Use of system dynamics and computer simulation in the system development.
c) Building of the smart sensor. (Examples are DSP-based velocity probes and accelerometers.)

Open Architecture with Mechatronic Models: Speed and Complexity [29]

Mechatronics plays the role irrespective of the possibility of single or multiple microcontrollers handling machine tools or an automobile assembly line of multiple robots. Simulating such complex systems allows designers to develop system without finalizing the hardware. The simulation procedure can be as "what if" scenario when the hardware does not exist. There are two critical issues to consider: speed and complexity. Larger systems involve more detailed simulation and specific system requirements. Tradeoffs between simulation speed and the level of accuracy are necessary because of system resources available. The simulation becomes faster with faster processors, but the use of multicore systems helps simulation. This is because the systems being simulated are distributed as well. Figure 6 shows an example of Stewart platform which is used in production lines and in many other industrial applications. The Stewart platform is a kind of parallel manipulator using an octahedral assembly of struts. A Stewart platform has six degrees of freedom (x, y, z, pitch, roll, and yaw). In this case, there are effectively two models: the simulated physical model and the application model. The physical model accounts for the physics-based simulated environment. The application model interacts with this environment to simulate the real-word application. Simulink and Matlab are used as model-based development tools, so the application is a model. The basic design represented in the physical world by computer-aided design and manufacturing tools (such as CATIA, Autodesk, and SolidWorks) has advanced simulation tools, although they are oriented toward physical construction rather than process control integration [29].

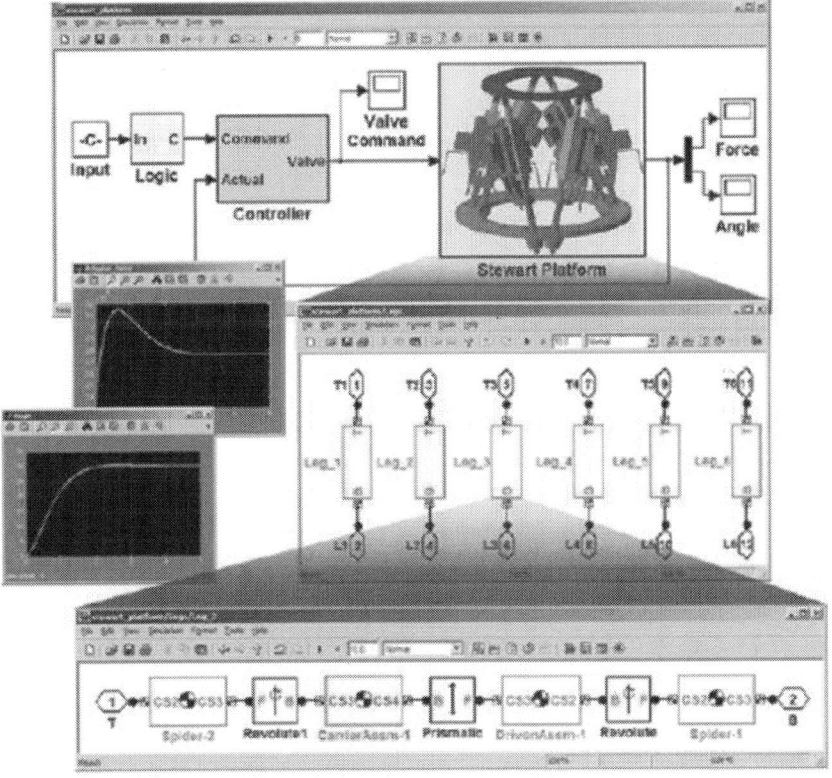

Figure 6. Simulink model of a platform [30].

The simulation platform can examine stress under a dynamic loading condition. It also addresses nonlinear analysis like deflection and impact with flexible materials such as foam, rubber, and plastic. In many cases, simulation and analysis of physical entities is useful in a design that does not include a computer-based controller. The contribution by National Instruments facilitates a major integration which facilitates the design engineers to bring in mechanical elements such as gears, cams, and actuators, while the programmers concentrate on the feedback and control algorithms that will handle the motors and actuators in the system. By linking various objects together we enable the models to interact. The provision of rendering permit the visualization of the models is in action. When creating large models, the modeling environment can demand significant amounts of computing power. The creation of large models can be a challenge to computing. At this stage, open architecture hosts can make a significant difference.

Multicore Speed Simulation [30]

Several CAD and model-based design systems employ interface software that takes advantage of multiple cores. New efforts are under way to develop and link several cores. Exploiting a large number of cores and clustered systems had been a challenge in advanced software architectures. The major challenge is communication between cores. The basic requirement of mechatronic simulation is the time synchronization between various objects in a distributed environment. Simulation in a multiple core environment is again a challenge when the shared memory cannot handle the synchronization. Typically there is an amount of limitation in the physical space. A robotic line-assembly simulation can perform well within its region, but will have a limited capability if it has to interact with other cells. Graphical model-based programming can assist in linking multiple cells (Figure 7).

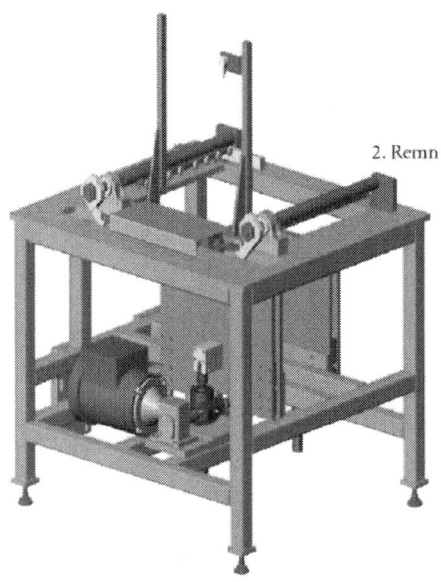

Figure 7. Assembly line design using "Cad" models [30].

Interactive Modeling

The key aspect of the graphical environments is that the visual representation of system partitioning and interaction lends itself to mechatronic applications. They also reduce system complexity from a developer's standpoint, allowing concentration on the application details. For example, the simulation tool such as Simscape is used as a declarative language that defines implicit relationships between components versus the explicit programming specifications for languages like C and C++ as well as graphical dataflow languages such as LabVIEW. Simscape targets cosimulation where programming and CAD intersect. This multidomain tool ties together the electronics, mechanical drive elements, mechanics, and hydraulics tools. For example, Stewart platform simulation discussed earlier can incorporate electrical, hydraulic, mechanical, and signal flow support in addition to software control of the system.

By reducing the amount of expertise required for developing mechatronic applications, developers can spend time and effort on other areas where they do have expertise. Likewise, having a model environment permits a better exchange of ideas and products. The difference these days is that the details within the models being exchanged as objects within a mechatronic application become more advanced. What used to be just dimensions is now something that can be used within a complete simulation with programmable feedback and even application interfaces when a model includes application code. The interactive modeling is crucial to the design process, and it can occur in a mixed environment where real and virtual objects are combined. A real robotic arm may be coupled with a virtual assembly line, for example, if the current task is to determine if the hand on the robotic arm can reorient an object. The robotic arm might be involved in laser welding of end plates. The key is getting the virtual objects and their control counterparts to interact with the real objects with a code that is running on remote devices. The electromechanical control systems once designed for the factory floor have become ubiquitous. For example, a designer may answer a problem of vibration by adding a stiffener. In an integrated mechatronic process, however, that small mechanical change may increase the mass of the part; it also may affect how fast the control system ramps up motor speed and how long the part holds in place before returning.

Figure 8 shows how a design verifier can assess whether an object is used correctly within the system. In this instance, the simulation verified the correct torque load under variable loading conditions. IN the availability of design verifier, assertion blocks are able to be included within a model so the system can determine whether an object's use within a system is correct.

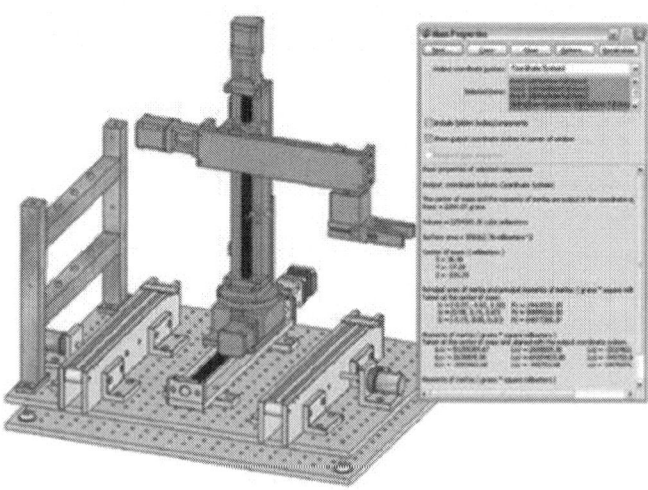

Figure 8. Simulation verification of torque load.

Many top mechatronic performers also use software that routes, tracks, and shares work. Most common are workflow tools, which automatically route work packages, warn about deadlines, and notify the right people of changes. Many companies make use of product data management tools to manage multidisciplinary bills of materials. Concurrent design is accomplished functionally using mechatronic approach. For example, engineers make assumptions about efficiency or how fast they can remove heat using certain construction techniques. Prototyping determines whether they hit their mark.

Right for the First Time: Virtual Machine Prototyping

The hardware in the loop facilitates the replacement of conventional mechanical motion control devices with digital devices. Mechanical systems are increasingly controlled by sophisticated electric motor drives that get their digital intelligence from software running on an embedded processor. Getting electromechanical designs right requires multidisciplinary teamwork and superb communication among team members. A decision like choosing the characteristics of a lead screw actuator has a ripple effect throughout the design and can impact the performance of other systems. To help facilitate a more integrated design process, we need to add motion simulation capabilities to CAD environments to create a more unified mechatronics workflow. Integrating motion simulation with CAD simplifies design because the simulation uses information that already exists in the CAD model, such as assembly mates, couplings, and material mass properties. Adding a high-level function block language for programming the motion profiles provides easier access to control those assemblies. This concept is known as "virtual machine prototyping" [31]. It brings together motion control software and simulation tools to create a virtual model of an electromechanical machine in operation. Virtual prototyping helps designers reduce risk by locating system-level problems, finding interdependencies, and evaluating performance tradeoffs. This is demonstrated in Figure 9 which shows the motion analysis for the green highlighted elements.

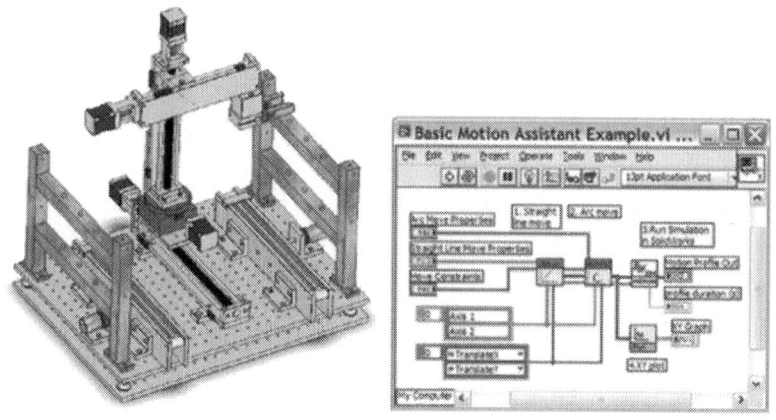

Figure 9. Evaluating "tradeoffs" in CAD environment [29].

Evaluating Trade off

Simulations enable everyone to work on development before the first prototype is completed. Engineers can use force and torque data from simulations for stress and strain analysis to validate whether mechanical components are stiff enough to handle the load during operation. They also can validate the entire operating cycle for the machine by driving the simulation with control system logic and timing. They can calculate a realistic estimate for cycle time performance, which is typically the top performance indicator for a machine design and compare force and torque data with the realistic limitations of transmission components and motors. This information can help identify flaws and drive design iterations from within the CAD environment. Simulations also simplify evaluating engineering tradeoffs between different conceptual designs. For example, would a SCARA robot be preferable to the 4-axis Cartesian gantry system? Simulations are faster and can be rerun whenever you make design changes. Consider an analysis of the torque load for the bottom lead screw actuator. If you violate the limits specified by the manufacturer, the mechanical transmission parts may not last for their rated life cycles. Using simulation software, you can find the mass of all the components mounted on the lead screw and determine the resulting center of mass by creating a reference coordinate system located at the center of the lead screw table and calculating the mass properties with respect to that coordinate system. With this information, you can calculate the static torque on the lead screw due to gravity caused by the overhanging load. Evaluating the dynamic torque induced by the motion is important because it tends to be much larger than the static torque load. Realistic motion profiles will help us to simulate inverse vehicle dynamics. This can provide more accurate torque and velocity requirements based on the motion profiles and the mass, friction, and gear ratio properties of the transmission.

Mechatronically Designed Ambulatory Rehabilitation Walker

The rehabilitation walker device designed and implemented by the authors in collaboration with a hospital is shown in Figure 10. This is an apparatus developed with the intent of aiding in the rehabilitation of hospital patients learning to walk again. This apparatus and control system are of industrial quality and would be reproducible in its entirety using off-the-shelf parts. The idea behind the rehabilitation walker is that it will relieve a certain percentage of body weight by carrying the patient in a harness which is attached to a hoist. The hoist is actively controlled using feedback from strain gauge sensors. As the patient walks around within the confines of the room-sized gantry, the hoist will follow the patient around. The overhead gantry is motorized in the X and Y directions. The closed loop motor control reacts to the feedback from multiaxis tilt sensors on the hoist line. If ever the patient were to fall, the hoist system would react and remove the full load of the patient's weight. The base of the control system consists of a National Instruments Compact RIO (reconfigurable input output) Programmable Automation Controller. The cRIO system is based on an FPGA backplane and a real-time controller. The backplane accepts modules which perform various I/O functions. The modules are chosen to interact with the rehabilitation walker sensors as well as handle the motor drive output signals. The motors are driven by industrial amplifiers, while position is tracked via quadrature encoder feedback. All programming of the system is done in LabVIEW (Figure 11).

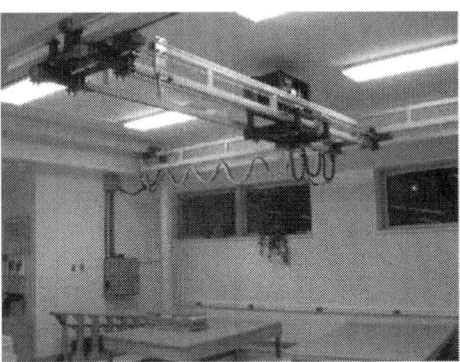
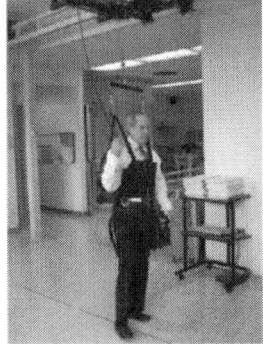

Figure 10. Mechatronic application for rehabilitation.

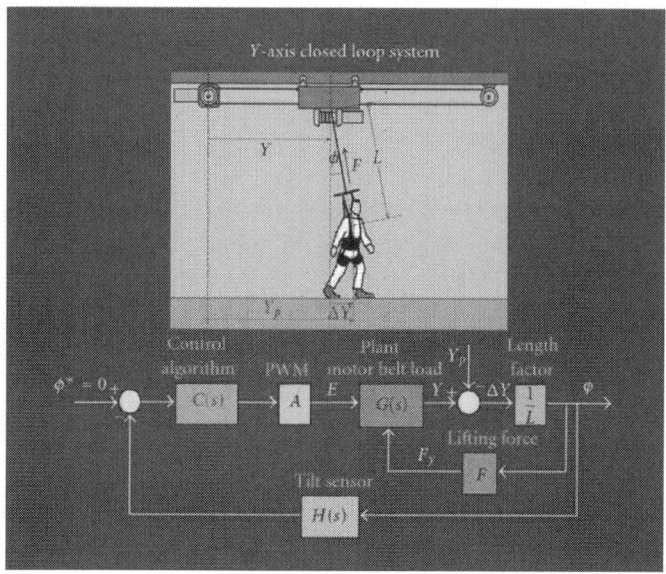

Figure 11. The Y-axis control loop system for the rehabilitation device "Navigator".

Evidence-Based Diagnostics

In healthcare field, Internet-based systems are available to help doctors identify possible causes for patient symptoms. One such statistical diagnostic assistant, called "Isabel," was developed by a father who sought to change the diagnostic system that affected the way his daughter (Isabel) was treated. This system is basically an intuitive system that takes advantage of all previous diagnoses and provides the statistically most likely disease (fault) and treatment (repair). The application of the condition-based maintenance information system is available in army and military applications. The system has the ability to integrate information from on-board sensors and diagnostic equipment to develop fleet wide logistic and situational awareness, implementing condition-based maintenance service that will enhance the operation and effectiveness of the tactical and combat vehicles.

Smart Machining Platform

The defense and aerospace industries are facing the challenges of quality and manufacturing efficiency. The US Army has sponsored a program called SMPI, which is organized into three modules. The first module consists of development of smart test bed for the integration, validation, and demonstration of SMPI technology areas. The second module deals with extension of test bed to find synergy between two or more thrust areas. The third module deals with integration of various thrust areas into a complete smart machine capable of making the first part correctly.

E-Manufacturing

E-manufacturing is a system methodology that enables the manufacturing operations to successfully integrate with the functional objectives of an enterprise through the use of Internet, tether-free (i.e., wireless, web, etc.), and predictive technologies [32, 33]. E-manufacturing includes the ability to monitor the plant floor assets, predict the variation and performance loss to dynamically reschedule production and maintenance operations, and synchronize related and consequent actions to achieve a complete integration between manufacturing systems and upper-level enterprise applications. The Rockwell automation report outlines a statement of competencies that are required of world class companies. These are design, operate, maintain, and synchronize. E-manufacturing should include intelligent maintenance and performance assessment systems to provide reliability, dependability, and minimum downtime with equipment running smoothly at their highest performance.

CONCLUSIONS

Increasing demands on the productivity of machine tools and their growing technological complexity call for improved methods in future product development processes. This paper examines the application of integrated tools for setting up a virtual prototype at early phases of the development process. Starting with mechatronics design process, the paper looks at open architecture issues, interactive modeling, and virtual prototyping. Mechatronics is also influenced by intelligent devices for the online and real-time monitoring which includes diagnosis and control of processes. Recent advances of mechatronics in smart manufacturing and modifications and improvements to conventional designs by using a mechatronics approach are discussed.

Products will continue to move toward greater complexity and an increasing integration of mechanical and electronic functions, including the growing trend toward ubiquitous computing and embedded systems. The way in which companies design and manufacture products has evolved rapidly with the emergence of global supply chains. Increasingly, components and subsystems will be sourced from worldwide suppliers and will need to come together seamlessly in subassembly and final assembly operations. There will be many more challenges as manufacturing locations are added and eliminated to meet contract demands and hold down import duties on new global contracts. Also, suppliers and components will be increasingly rotated in and out of a design to meet the challenges of global supply chains and cost reduction pressures. Manufacturing will be far more fluid and will need to adjust rapidly to new locations, component suppliers, and design changes. For all of these reasons, advanced mechatronic tools will be required to assure high-quality manufacturing from the design for manufacture to the continuous monitoring and modification of the production processes. They will be essential for both high quality and profitability.

REFERENCES

1. Y. Li, J. Huang, and H. Tang, "A compliant parallel XY micro-motion stage with complete kinematic decoupling," IEEE Transactions on Automation Science and Engineering, vol. 9, no. 3, pp. 538–553, 2012.
2. Y. Yun and Y. Li, "Optimal design of a 3-PUPU parallel robot with compliant hinges for micromanipulation in a cubic workspace," Robotics and Computer-Integrated Manufacturing, vol. 27, no. 6, pp. 977–985, 2011.
3. Y. Tang and Y. Li, "The software architecture of a reconfigurable real-time onboard control system for a small UAV helicopter," in Proceedings of the 8th International Conference on Ubiquitous Robots & Ambient Intelligence (URAI '11), pp. 228–233, Songdo Convention, Incheon, Korea, November 2011.
4. Y. Tang and Y. Li, "Development of a laboratory HILs testbed system for small UAV helicopters," in Proceedings of the IASTED International Conference on Robotics (Robo '11), pp. 428–436, Pittsburgh, Pa, USA, November 2011.
5. J.-G. Wang and Y. Li, "Hybrid impedance control of a 3-DOF robotic arm used for rehabilitation treatment," in Proceedings of the 6th IEEE International Conference on Automation Science and Engineering (CASE '10), pp. 768–773, Ontario, Canada, August 2010.
6. D. Shetty and R. Kolk, Mechatronic System Design, Thomson Engineering Publications, Toronto, Canada, 1998.
7. G. Reinhart and M. Weissenberger, "Multibody simulation of machine tools as mechatronic systems for optimization of motion dynamics in the design process," in Proceedings of the IEEE/ASME International Conference on

Advanced Intelligent Mechatronics (AIM '99), pp. 605–610, Atlanta, Ga, USA, September 1999.
8. D. K. Miu, Mechatronics, Springer, Berlin, Germany, 1993.
9. G. Reinhart, M. WeiBenberger, A. Sprenzel, J. Meinlschmidt, and P. Wagner, "Innovative entwicklung von werkzeugmaschinen -methoden und entwicklungswerkzeuge," in Synergy of Culture and Production, Vol. 1- Holistic Approach To Machine Tool Innovation, H. Y. Ito and E. F. Moritz, Eds., Artefact, Sottrum, Germany, 1997.
10. K. Erkorkmaz and Y. Altintas, "High speed contouring control algorithm for CNC machine tools," inProceedings of the ASME International Mechanical Engineering Congress and Exposition, pp. 463–469, November 1998.
11. H. S. Cho and M. Y. Kim, "Optomechatronic technology: the characteristics and perspectives," IEEE Transactions on Industrial Electronics, vol. 52, no. 4, pp. 932–943, 2005.
12. K. Kincade, "Atomic-force microscopy finds new role in the nano world," Laser Focus World, vol. 40, no. 4, pp. 87–91, 2004.
13. N. Wakami, H. Nomura, and S. Araki, "Fuzzy logic for home appliances," in Fuzzy Logic and Neural Networks, C. H. Chen, Ed., pp. 21.1–21.23, McGraw- Hill, New York, NY, USA, 1996.
14. T. J. Krupa, "Optical R&D at the army research laboratory," Optics and Photonics News, vol. 11, no. 6, pp. 16–25, 2000.
15. S. D. Robinson, "MEMS technology—micromachines enabling the all optical network," in Proceedings of the 51st Electronic Components and Technology Conference, pp. 423–428, Orlando, Fla, USA, June 2001.
16. K. Tsuruta, Y. Mikuriya, and Y. Ishikawa, "Micro sensor developments in Japan," Sensor Review, vol. 19, no. 1, pp. 37–42, 1999.
17. T. V. Higgins, "Optical storage lights the multimedia future," Laser Focus World, vol. 31, no. 9, p. 103, 1995.
18. S. G. Anderson, "Smart cars take the high-tech road," Laser Focus World, vol. 32, no. 6, p. 117, 1996.
19. T. G. McDonald and L. A. Yoder, "Digital micromirror devices make projection displays," Laser Focus World, vol. 33, no. 8, pp. S5–S8, 1997.
20. H. S. Cho, Opto-Mechatronic Systems Handbook: Techniques and Applications, CRC Press, Boca Raton, Fla, USA, 2003.
21. D. A. Bradley, D. Dawson, N. C. Burd, and A. J. Loader, Mechatronics: Electronics in Products and Processes, Chapman & Hall, London, UK, 1991.
22. Y. Altintas, C. Brecher, M. Week, and S. Witt, "Virtual machine tool," CIRP Annals, vol. 54, no. 2, pp. 651–674, 2005.
23. H. Van Brussel, P. Sas, I. Németh, P. De Fonseca, and P. Van Den Braembussche, "Towards a mechatronic compiler," IEEE/ASME Transactions on Mechatronics, vol. 6, no. 1, pp. 90–105, 2001.

24. M. F. Zaeh, T. Oertli, and J. Milberg, "Finite element modelling of ball screw feed drive systems," CIRP Annals, vol. 53, no. 1, pp. 289–293, 2004.
25. G. Pritschow, "On the influence of the velocity gain factor on the path deviation," CIRP Annals, vol. 45, no. 1, pp. 367–371, 1996.
26. Y. Altintas and M. Weck, "Chatter stability of metal cutting and grinding," CIRP Annals, vol. 53, no. 2, pp. 619–642, 2004.
27. R. Gurbuz, "Mechatronics approach for desk-top CNC milling machine design," Diffusion and Defect Data B, vol. 144, pp. 175–180, 2009.
28. Siemens PLM Software, CHICAGO (IMTS), 2010, http://www.siemens.com/industryautomation.
29. B. M. Cleery and N. Mathur, "Right the first time," Mechanical Engineering, vol. 130, no. 6, pp. 30–34, 2008.
30. W. Wong, Muticore Matters with Mechatronic Models Electronic Design, 2008.
31. NI LabVIEW-Solid Works mechatronics toolkit, http://www.ni.com/mechatronics/.
32. A. Ali, Z. Chen, and J. Lee, "Web-enabled platform for distributed and dynamic decision-making systems," International Journal of Advanced Manufacturing Technology, vol. 38, no. 11-12, pp. 1260–1270, 2008.
33. J. Lee, "E-manufacturing-fundamental, tools, and transformation," Robotics and Computer-Integrated Manufacturing, vol. 19, no. 6, pp. 501–507, 2003.

CITATION

Devdas Shetty, Lou Manzione, and Ahad Ali, "Survey of Mechatronic Techniques in Modern Machine Design," Journal of Robotics, vol. 2012, Article ID 932305, 9 pages, 2012. doi:10.1155/2012/932305.

CHAPTER 2

Method for Remanufacturing Large-Sized Skew Bevel Gears Using CNC Machining Center

Kazumasa Kawasaki[1], Isamu Tsuji[2], Hiroshi Gunbara[3], Haruo Houjoh[4]

[1]Institute for Research Collaboration and Promotion, Niigata University, 8050 Ikarashi 2-nocho, Nishi-ku, Niigata 950-2181, Japan
[2]Iwasa Tech Co., Ltd., Takasecho 64-2, Funabashi 273-0014, Japan
[3]Department of Mechanical Engineering, Matsue National College of Technology, 14-4, Nishi-ikuma-cho, Matsue 690-8518, Japan
[4]Precision & Intelligence Laboratory, Tokyo Institute of Technology, 14-4, Nagatsuta 4259, Midori-ku, Yokohama 226-8503, Japan

HIGHLIGHTS

- A method for remanufacturing large-sized skew bevel gears using a CNC machining center is proposed.
- The deviations between the real and theoretical tooth surface forms of gear member are formalized.
- The tooth contact pattern and transmission errors reflecting the formalized deviations are analyzed.
- The pinion and gear members of skew bevel gears are remanufactured by swarf cutting.
- The pinion member that has good performance mating with an existing gear member can be machined.

ABSTRACT

A method for remanufacturing pinion member of large-sized skew bevel gears using a CNC machining center and respecting an existing gear member is proposed. For this study, first the tooth surface forms of skew bevel gears are modeled mathematically. Next, the real tooth surfaces of the existing gear member are measured using a coordinate measuring machine and the deviations between the real and theoretical tooth surface forms are formalized using polynomial expression. Moreover, the tooth contact pattern and transmission errors reflecting the deviations of the tooth surface forms of the gear member are analyzed, and the tooth surface form of the pinion member that has good performance mating with the existing gear member is designed. The pinion member was remanufactured by swarf cutting using a CNC machining center. The gear member is also remanufactured using this method in order to apply to the case where pinion member exists in reverse or both gear and pinion members do not exist. The tooth surface form deviations were detected, and the experimental tooth contact pattern of the pinion and gear members was compared with analytical one. The results showed good agreement.

INTRODUCTION

Bevel gears are used to transmit power and motion between the intersecting axes of the two shafts, and are most often mounted on shafts that are 90° apart. There are four basic types of bevel gears that may have straight, Zerol (by Greason Works), spiral, and skew teeth [1], [2], [3] and [4], and occupy an important place in gear transmissions [5]. Almost all bevel gears are manufactured using a special generator, they are commonly used in pairs, and thus usually are not interchangeable. In such case, the performance of a gear pair is usually evaluated by tooth contact pattern.

The transmission of straight bevel gears is regarded as a particular case of skew bevel gears [6]. The contact ratio of skew gears is larger than that of straight bevel gears because skew bevel gears have oblique teeth. Such skew bevel gears are used at power generation plants when the gears are large in size.

INTRODUCTION

It is now possible to machine the complicated tooth surface using a multipurpose machine due to the development of multi-axis control and multi-tasking machine tools[7] and [8]. Therefore, high precision machining of large-sized skew bevel gears has been expected. The manufacturing method of some bevel gears has been proposed by making use of this development process [9], [10], [11] and [12]. The manufacturing method has the advantages of arbitrary modification of the tooth surface and machining of the part except the tooth surface [12].

Meanwhile, some methods for the optimization of the gear contact on the correction of the parameter of the special machine tool in spiral bevel and hypoid gear cutting have been proposed [13], [14], [15], [16] and [17]. With the exception of a few approaches [17], both the gear and pinion members are usually modified.

In recent years, the renovation of power generation plants has been active due to the age of the plants. In order to advance this renovation, it has become necessary to replace the skew bevel gears in the plants. In this situation, the pair has to be replaced even if the gear member can still be used. However, due to its large size, the special machines do not exist in present. Moreover, the tool wear, the long operating time of the machine etc. occur. These lead to high costs in gear production. Therefore, it is natural to replace only the pinion member. In this case, it is necessary to remanufacture the pinion member that has good performance mating with an existing gear member.

The authors have proposed a manufacturing method of the pinion member of the large-sized skew bevel gears using a CNC machining center and respecting an existing gear member [18]. However, there are the obscure parts in design, manufacturing, and evaluation of the pinion member mating with an existing gear member. In this paper, the remanufacturing of the pinion member is completed. Moreover, the gear member is also remanufactured using this method in order to apply to the case where only the pinion member exists in reverse or both gear and pinion members do not exist.

The approach of this paper involves the following six steps.

1) The tooth surface forms of skew bevel gears are modeled mathematically and simply.
2) The real tooth surfaces of the existing gear member are measured using a coordinate measuring machine (CMM) and the deviations between the real and theoretical tooth surface forms are formalized as polynomial expression using the measured coordinates.
3) The tooth contact pattern and transmission errors reflecting the deviations of the tooth surface forms of the gear member are analyzed, and the tooth surface form of the pinion member that has good performance mating with the existing gear member is designed.
4) The pinion member is remanufactured by swarf cutting that is machined by the side of the end mill using a CNC machining center.
5) The real tooth surfaces of the remanufactured pinion member are measured using a CMM and the tooth surface form deviations are detected. Moreover, the tooth contact patterns of the remanufactured pinion member and existing gear member are investigated.
6) The gear member with the tooth surface form of the existing gear member is also remanufactured using this method and the tooth surface form deviations are detected. Moreover, the tooth contact patterns of the remanufactured pinion and gear members are investigated.

As a result, the tooth surface form deviations of the remanufactured pinion and gear members were a permitted limit on both the drive and coast sides, respectively. The experimental tooth contact patterns showed good agreement compared with analytical one.

TOOTH SURFACES OF SKEW BEVEL GEARS

In this section, the tooth surface forms of skew bevel gears are modeled mathematically and simply in order to obtain three-dimensional coordinates and unit normals on the tooth surface considering the machining using a CNC machining center.

In general, the geometry of the skew bevel gears is achieved by considering an imaginary crown gear as the theoretical generating tool.

Therefore, first the tooth surface form of the imaginary crown gear is considered.

The number of teeth of the imaginary crown gear is represented by

$$z_i = \frac{z_p}{\sin \lambda_{po}} = \frac{z_g}{\sin \lambda_{go}} \tag{1}$$

where z_p and z_g are the number of teeth of pinion and gear, respectively, and λ_{po} and λ_{go} are the pitch cone angles of the pinion and gear, respectively.

Fig. 1 shows the tooth surface form of the imaginary crown gear assuming to be straight bevel gears with depthwise tooth taper. O-xyz is the coordinate system fixed to the crown gear and z axis is the crown gear axis of rotation. Point P is a reference point at which tooth surfaces mesh with each other and is defined in the center of tooth surface. R_m is the mean cone distance. b is the face width. M_n is the normal module. α is the pressure angle. The circular arcs with large radii of curvatures are defined both in xz and xy planes. xz and xy planes correspond to the sections of the tooth profile and tooth trace of the tooth surface, respectively. This curved surface is defined as the tooth surface of the imaginary crown gear. u and θ are the parameters which represent curved lines. Δc and Δs are the amounts of tooth profile modification and crowning, respectively. ρ_c and ρ_s are the radii of curvatures of circular arcs in xz and xy planes, respectively. ρ_c and ρ_s have influence on Δc and Δs, respectively. The following equations yield considering the relations between ρ_c, Δc, M_n, and α in xz, and between ρ_s, Δs, and b in xy planes, respectively [19]:

$$\rho_c = \frac{\Delta c^2 + \left(M_n/\cos\alpha\right)^2}{2\Delta c}$$

$$\rho_s = \frac{\Delta s^2 + \dfrac{b^2}{4}}{2\Delta s}, \qquad (2)$$

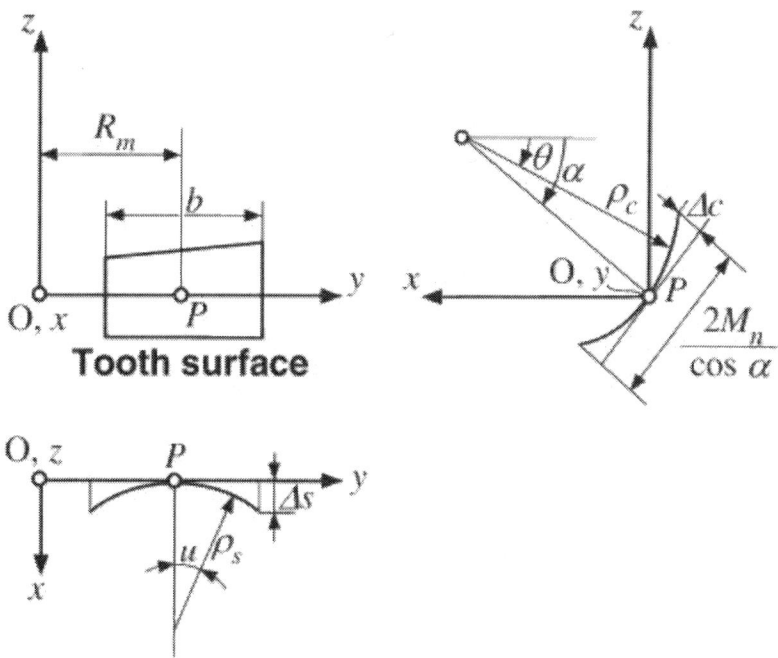

Figure 1. Tooth surface form of imaginary crown gear.

Since skew bevel gears have teeth that are straight and oblique, the skew bevel gears have the skew angle. Therefore, the imaginary crown gear also has the skew angle that is defined as β as shown in Fig. 2. The tooth surface of the imaginary crown gear is expressed in O-xyz using ρ_c and ρ_s:

$$X(u, \theta) = \begin{bmatrix} -\rho_c(\cos\theta - \cos\alpha) - \rho_s(1 - \cos u) + \rho_s \sin u \tan\beta \\ \rho_s \sin u + R_m \\ \rho_c(\sin\alpha - \sin\theta) \end{bmatrix}.$$

(3)

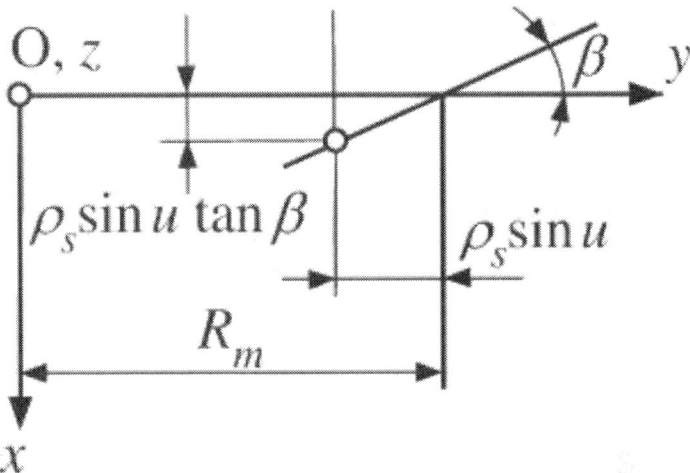

Figure 2. Skew angle of imaginary crown gear.

The unit normal of X is expressed by N.

X expresses the equation of the tooth surface of the imaginary crown gear. The imaginary crown gear is rotated about the z axis by angle ψ and generates the tooth surface of the skew bevel gear. This rotation angle, ψ, of the crown gear, is the generating angle. When the generating angle is ψ, X and N are rewritten as X_ψ and N_ψ in O-$x_s y_s z_s$ assuming that the coordinate system O-xyz is rotated about the z axis by ψ in the coordinate system O-$x_s y_s z_s$ fixed in space. When ψ is zero, O-$x_s y_s z_s$ coincides with O-xyz.

Assuming the relative velocity $W(X_\psi)$ between the crown gear and the generated gear at the moment when generating angle is ψ, the equation of meshing between the two gears is as follows [20] and [21]:

$$N_\psi(u, \theta; \psi) \cdot W(u, \theta; \psi) = 0. \qquad (4)$$

From Eq. (4), we have $\theta = \theta(u, \psi)$. Substituting $\theta(u, \psi)$ into X_ψ and N_ψ, any point on the tooth surface of the crown gear and its unit normal are defined by a combination of (u, ψ), respectively. When the tooth surface of the imaginary crown gear in O-$x_s y_s z_s$ is transformed into the coordinate system fixed to the generated gear, the tooth surface of the skew bevel gear is expressed. The theoretical tooth surfaces of the pinion and gear are expressed as x_p and x_g, respectively. Moreover, the unit normals of x_p and x_g are expressed as n_p and n_g, respectively. Henceforth, the subscripts "p" and "g" indicate that each is related to the pinion and gear, respectively.

MEASUREMENT OF EXISTING GEAR MEMBER

The detail of the manufacturing method of the existing gear member is uncertain. Moreover, manufacturing errors occur in bevel gear cutting. Therefore, whether the mathematical model as mentioned in Section 2 fits the real tooth surface of the existing gear member or not is not clear. In this section, the tooth surfaces of the gear member are measured using a CMM and the deviations between the real and theoretical tooth surface forms are formalized as polynomial expression.

Coordinate measurement of real tooth surface
The theoretical tooth surface of the existing gear member is expressed as $x_g(u_g, \psi_g)$ as mentioned in Section 2. A grid of n lines and m columns is defined and a point called the reference point is specified on the tooth surfaces of drive and coast sides, respectively. The reference point is usually chosen in the center of the grid. The position vector $x_g(x, y, z)$, namely, u, θ, and ψ is determined for the solution of simultaneous equations considering one point on the grid of the tooth surface and the unit normal (n_x, n_y, n_z) of the corresponding surface point is also determined since u, θ, and ψ are determined [22].

For measurement, the existing gear member is set up arbitrarily on a CMM whose coordinate system is defined as O_m-$x_m y_m z_m$. We can make origin O_m and axis z_m coincide with the origin and the axis of the gear member, respectively. The whole grid of surface points together with the theoretical tooth surfaces is rotated about z_m axis so that y_m is equal to zero at the reference point. Therefore, the position vector of the point and its unit normal are transformed into the coordinate system O_m-$x_m y_m z_m$ and are represented by

$$\begin{aligned} \mathbf{x}^{(i)} &= \left(x^{(i)}, y^{(i)}, z^{(i)}\right)^T \\ \mathbf{n}^{(i)} &= \left(n_x^{(i)}, n_y^{(i)}, n_z^{(i)}\right)^T \end{aligned} \quad (i = 1, 2, \cdots, 2n \times m) \tag{5}$$

$\mathbf{x}^{(i)}$ and $\mathbf{n}^{(i)}$ in Eq. (5) are the i-th coordinates of the theoretical tooth surface and its unit normal, respectively.

The real tooth surface of the existing gear member was measured using a CMM (Sigma M & M 3000 developed by Gleason Works). When the real tooth surface is measured according to the provided grid, the i-th measured tooth surface coordinates are obtained and are numerically expressed as the position vector [22] and [23]:

$$\mathbf{x}_m^{(i)} = \left(x_m^{(i)}, y_m^{(i)}, z_m^{(i)}\right)^T \quad (i = 1, 2, \cdots, 2n \times m). \tag{6}$$

When the deviation δ between the measured coordinates and the coordinates of theoretical tooth surfaces for each point on the grid is defined toward the direction of normal of theoretical tooth surface, i-th δ can be determined by

$$\delta^{(i)} = \left(\mathbf{x}_m^{(i)} - \mathbf{x}^{(i)}\right) \cdot \mathbf{n}^{(i)} \quad (i = 1, 2, \cdots, 2n \times m) \tag{7}$$

where δ is equal to zero at the reference point. The fundamental components of polynomial expression formalizing the deviations of tooth surface forms are used because the motion concept may be implemented on a CNC machining center.

Formalization of deviations of tooth surface form

Based on the method as mentioned in Section 3.1, the deviation δ for each point on the grid is calculated when the points on the real tooth surface are measured [24]. However, it is difficult to fit δ to the expressed theoretical tooth surface well because δ varies at each point on the grid. The fourth-order and sixth-order functions have been proposed in order to formalize the deviations of the tooth surface form of spiral bevel and hypoid gears [15],[23] and [25]. The goal of this paper is to manufacture the pinion member that has good performance mating with the existing gear member using a CNC machining center and to obtain a good tooth contact pattern of a gear pair. The target value of the deviation is decided as values less than about 0.05–0.06 mm focusing the position of tooth contact pattern of the slew bevel gears of the gear member with large pitch circle diameter of more than 1000 mm. Therefore, the formalization of the deviations is simplified and the third order functions are utilized for convenience.

We define (X, Y) whose X and Y are toward the directions of the tooth profile and tooth trace, respectively, and form the following polynomial expression:

$$\Delta = \delta_{11} + \delta_{12} + \delta_{21} + \delta_{22} + \delta_{31} + \delta_{32} + \delta_{41} \tag{8}$$

where $\delta_{11}, \delta_{12}, \delta_{21}, \delta_{22}, \delta_{31}, \delta_{32}$, and δ_{41} are defined as follows: Fig. 3 shows the procedure formalizing the relation between the fundamental components of polynomial expression and the deviation of tooth surface form. First, the tooth trace deviation δ_{11} and tooth profile deviation δ_{12} are expressed as the following first order equations of X and Y using fundamental components a_{11} and a_{12}, respectively [see Fig. 3(a)]:

$$\delta_{11} = a_{11}X$$
$$a_{11} = \frac{\delta_{11}}{0.5H}$$
$$\delta_{12} = a_{12}Y$$
$$a_{12} = \frac{\delta_{12}}{0.5T} \qquad (9)$$

where H and T are defined as the ranges of the evaluation of the tooth surface in X and Y directions, respectively. Next, the tooth trace deviation δ_{21} and tooth profile deviation δ_{22} are expressed as the following second order equations of both X and Y using fundamental components a_{21} and a_{22}, respectively [see Fig. 3(b)]:

$$\delta_{21} = a_{21}X^2$$
$$a_{21} = \frac{\delta_{21}}{(0.5H)^2} = \frac{4\delta_{21}}{H^2}$$
$$\delta_{22} = a_{21}Y^2$$
$$a_{22} = \frac{\delta_{22}}{(0.5T)^2} = \frac{4\delta_{22}}{T^2} \qquad (10)$$

where δ can be calculated in consideration of only (a) and (b) in Fig. 3 by the following equation:

$$\delta = \delta_{11} + \delta_{12} + \delta_{21} + \delta_{22}$$
$$= a_{11}X + a_{12}Y + a_{21}X^2 + a_{22}Y^2. \qquad (11)$$

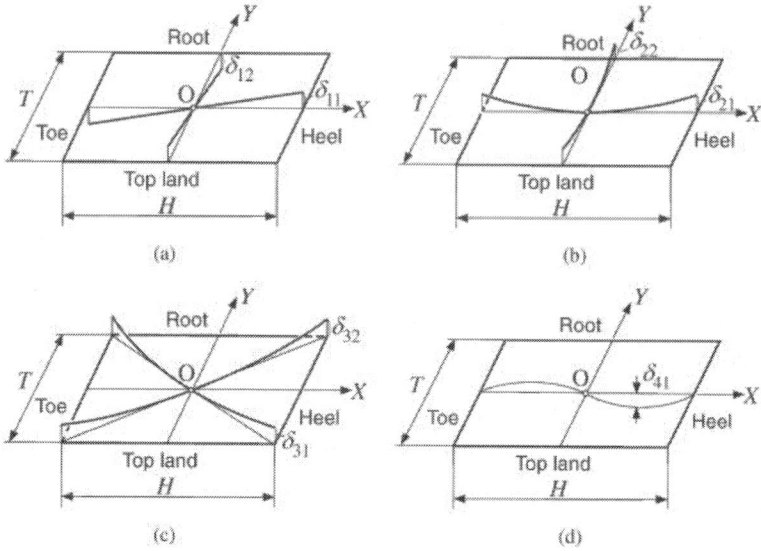

Figure 3. Procedure formalizing relation between fundamental components of polynomial expression and deviation of tooth surface form.

The deviations may occur in the directions of the bias-in and bias-out according to δ in Eq. (11). Therefore, the deviations δ_{31} and δ_{32} in the directions of the bias-in and bias-out are expressed as the following second order equations of both X and Y using fundamental components a_{31} and a_{32}, respectively [see Fig. 3(c)]:

$$\xi_1 = \tan^{-1}\left(\frac{T}{H}\right)$$

$$L_0 = \frac{H}{\cos\xi_1}$$

$$\delta_{31} = a_{31}(X\cos\xi_1 - Y\sin\xi_1)^2$$

$$a_{31} = \frac{\delta_{31}}{(0.5L_0)^2} = \frac{4\delta_{31}}{L_0^2}$$

$$\delta_{32} = a_{32}(X\cos\xi_1 + Y\sin\xi_1)^2$$

$$a_{32} = \frac{\delta_{32}}{(0.5L_0)^2} = \frac{4\delta_{31}}{L_0^2}.$$

(12)

Finally, the tooth trace deviation δ_{41} is expressed as the following third order equations of X and Y using fundamental components b_1, b_2, and b_3, respectively [see Fig. 3(d)]:

$$\delta_{41} = b_3 X^3 + b_2 X^2 + b_1 X \qquad (13)$$

where b_1, b_2, and b_3 are determined from the following conditions: δ is equal to zero when $X = -0.5H$ and $X = 0.5H$. Moreover, δ is equal to δ_{41} when $X = 0.25H$. The deviation δ is calculated in consideration of Eq. (12) by the following equation:

$$\delta = A_1 X + A_2 Y + A_3 X^2 + A_4 Y^2 + A_5 XY + A_6 X^3 \qquad (14)$$

where

$$A_1 = a_{11} + b_1$$
$$A_2 = a_{12}$$
$$A_3 = (a_{31} + a_{32})\cos^2 \xi_1 + a_{21} + b_2$$
$$A_4 = (a_{31} + a_{32})\sin^2 \xi_1 + a_{22}$$
$$A_5 = 2(a_{32} - a_{31})\cos\xi_1 \sin\xi_1$$
$$A_6 = b_3. \qquad (15)$$

Reflecting the polynomial expression δ to the theoretical tooth surface, the position vector is represented by

$$\mathbf{x}_a = \mathbf{x} + \delta \mathbf{n} \qquad (16)$$

where $\mathbf{x}_a$ represents the theoretical tooth surface in consideration of the tooth surface form deviations. The tooth contact patterns and transmission errors with the tooth surface form deviations are analyzed using $\mathbf{x}_a$, that is, the numerical coordinates obtained using $\mathbf{x}_a$ are considered as nominal data of the existing gear member. The position vector of i-th point of the theoretical tooth surface is expressed as $\mathbf{x}_a^{(i)}$.

Measured results and formalization of deviations

The real tooth surfaces of the existing gear member to be used were measured using a CMM as shown in Fig. 4 and the deviations between the real and theoretical tooth surface forms were formalized. Table 1 shows the dimensions of the skew bevel gears. The pitch circle diameter of the gear member is 1702.13 mm and it is very large. Fig. 5shows the required tooth contact pattern. The position of the tooth contact pattern is somewhat near the toe side and the length is about 50% of the tooth width. Five points in the direction of the tooth profile and nine points in the direction of the tooth trace for the grid were used. The amounts of tooth profile modification and crowning are $\Delta c = 0.05$ mm and $\Delta s = 0.05$ mm. Fig. 6 shows the formalized results based on the measured coordinates. Fig. 6(a) shows the measured results using a CMM without formalization of the deviations δ between the real and theoretical tooth surface forms. Fig. 6(b) shows the formalized results using δ in Eq. (11). In this case, the deviations in the directions of the bias-in and bias-out are not considered. Table 2 shows the calculated results of coefficients a_{11}, a_{12}, a_{21}, and a_{22} of polynomials corresponding to Fig. 6(b). Fig. 6(c) shows the formalized results using δ in Eq. (14). In this case, the deviations in the directions of the bias-in and bias-out are considered. Table 3 shows the calculated results of coefficients A_1, A_2, A_3, A_4, A_5, and A_6 of polynomials corresponding to Fig. 6(c).

Figure 4. Measurement of gear member using CMM.

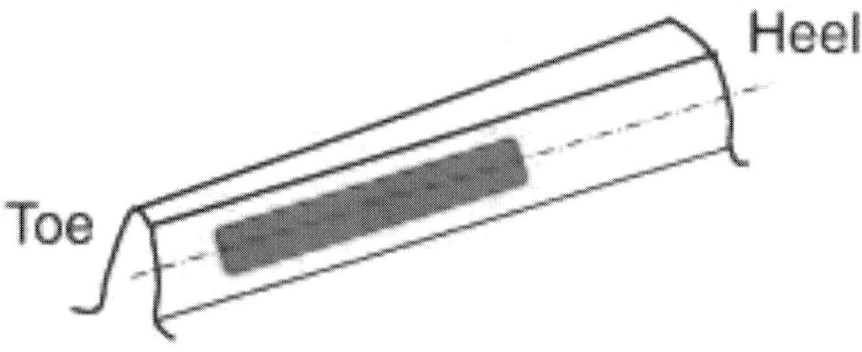

Figure 5. Required tooth contact pattern.

Table 1. Dimensions of skew bevel gears.

		Pinion	Gear
Number of teeth	z_p, z_g	18	116
Pitch circle diameter		264.1346 mm	1702.1302 mm
Pitch cone angle		8.8167 deg.	81.1833 deg.
Normal module M_n		10.6764	
Mean cone distance R_m		759.65 mm	
Pressure angle α		14.5 deg.	
Skew angle β		15 deg.	
Face width b		203.2 mm	
Shaft angle		90 deg.	
Backlash		0.4064–0.5588 mm	

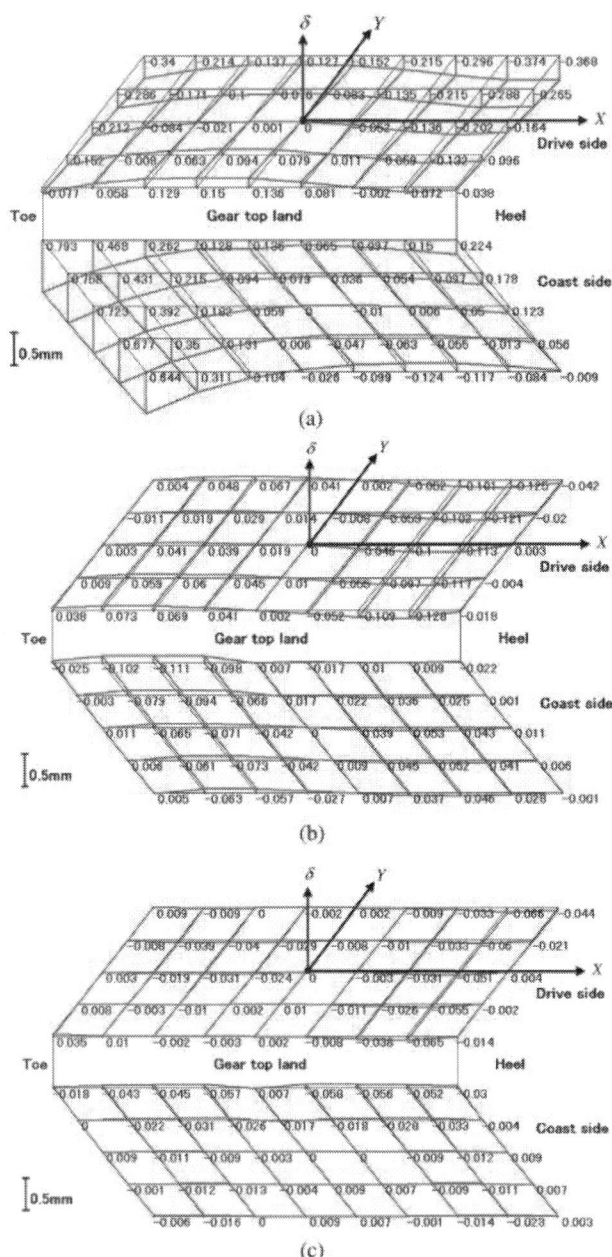

Figure 6. Formalized results based on measured coordinates [mm].

Table 2. Calculated results of coefficients of polynomial using δ in Eq. (11).

	a_{11}	a_{12}	a_{21}	a_{22}
Drive side	18.75876	-0.13576	0.14046	-0.0224
Coast side	15.30663	0.31395	-3.37429	0.0504

Table 3. Calculated results of coefficients of polynomial using δ in Eq. (14).

	A_1	A_2	A_3	A_4	A_5	A_6
Drive side	18.7588	-0.1736	-1.8828	-0.0232	0.0715	0.00024
Coast side	15.3066	0.19594	-1.5693	0.04877	0.05805	-0.0002

The maximum values of the magnitude of deviations are 0.793 mm, 0.128 mm, and 0.066 mm in Fig. 6(a), (b), and (c), respectively. The overall deviations gradually decrease as a whole from Fig. 6(a) to Fig. 6(c). Since the deviations are formalized with a large number of equations as mentioned in Section 3.2, the deviations decrease and fit the measured coordinates well to the expressed theoretical tooth surface. Moreover, the value seems to be acceptable, although the maximum value 0.066 mm of the magnitude of deviation in Fig. 6(c) is larger somewhat than the target value. Therefore, the validity of the formalization of the deviations was confirmed.

TOOTH CONTACT ANALYSIS

Concept of tooth contact analysis

The tooth surface form of the pinion member that has good performance mating with the existing gear member is considered based on tooth contact analysis. In this case, the tooth surface form of the pinion member is designed and the appropriate amounts of profile modification and crowning are calculated.

The pinion and gear members are assembled in a coordinate system O_h-$x_h y_h z_h$ as shown in Fig. 7 in order to analyze the tooth contact pattern and transmission errors of the pinion and gear members. Suppose that ϕ_p and ϕ_g are the rotation angles of the pinion and gear, respectively. The position vectors of the pinion and gear tooth surfaces must coincide and the direction of two unit normals at this position must be also coincide in order to contact the two surfaces. Therefore, the following equations yield [25]:

$$B(\phi_p)x_p(u_p,\psi_p) = C(\phi_g)x_g(u_g,\psi_g)$$
$$B(\phi_p)n_p(u_p,\psi_p) = \pm C(\phi_g)n_g(u_g,\psi_g) \qquad (17)$$

where **B** and **C** are the coordinate transformation matrices for the rotation about y_h and z_h axes, respectively:

$$B(\phi_p) = \begin{bmatrix} \cos\phi_p & 0 & \sin\phi_p \\ 0 & 1 & 0 \\ -\sin\phi_p & 0 & \cos\phi_p \end{bmatrix}$$

$$C(\phi_g) = \begin{bmatrix} \cos\phi_g & -\sin\phi_g & 0 \\ \sin\phi_g & \cos\phi_g & 0 \\ 0 & 0 & 1 \end{bmatrix}. \qquad (18)$$

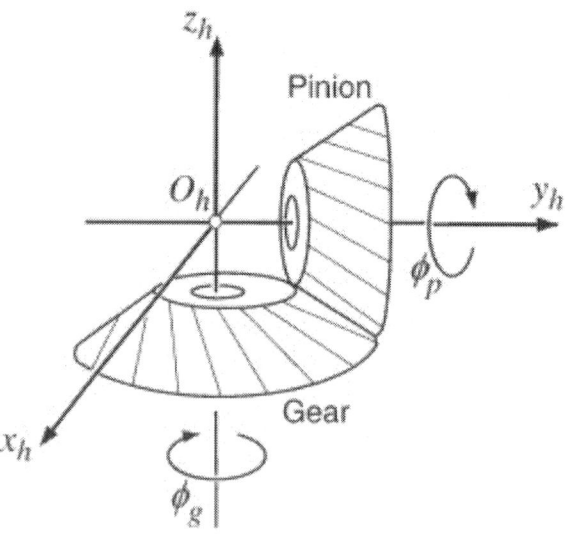

Figure 7. Meshing of pinion and gear.

Since $|n_p| = |n_g| = 1$, Eq. (17) represents a system of five scalar non-linear equations with five unknowns u_p, ψ_p, u_g, ψ_g, and ϕ_g considering angle ϕ_p as the input parameter. The continuous solution of the system of the non-linear equations permits the determination of the path of contact considering that ϕ_p changes every moment. A method of successive approximation is utilized in order to obtain a numerical solution of Eq. (17). In this case, it is convenient to use a cylindrical coordinate system.

The paths of contact on the pinion and gear tooth surfaces are represented by $x_p(u_p, \psi_p)$ and $x_g(u_g, \psi_g)$, respectively.

When the pinion is rotated by the angle ϕ_p, the gear should be rotated by the angle $z_p/z_g \cdot \phi_p$ assuming that the pinion and gear are conjugate. In practice, however, this is not the case, and there are transmission errors. The function of transmission errors is defined as:

$$\Delta\phi_g(\phi_p) = \phi_g(\phi_p) - \frac{z_p}{z_g}\phi_p. \qquad (19)$$

Results of tooth contact analysis

The tooth contact pattern and transmission errors of the pinion and gear members whose tooth surfaces were determined were analyzed under unloaded condition based on the method as mentioned in Section 4.1.

Fig. 8 shows the analyzed results of tooth contact pattern and transmission errors without taking into account tooth surface form deviations δ as mentioned in Section 3.2. Fig. 8(a) is the result of the drive side and Fig. 8(b) is that of the coast side. The amounts of tooth profile modification and crowning of pinion member are $\Delta c = 0.05$ mm and $\Delta s = 0.05$ mm, on both drive and coast sides, respectively. These values have influence on the tooth contact pattern and transmission errors. The right side from the center in Fig. 8 shows the analyzed contour line on the gear tooth surface at every instant when the rotation angle ϕ_p of the pinion changes from 12.5° to − 27.5° on the drive side and from − 12.5° to 27.5° on the coast side, respectively. The lowest figure of the right side shows the total tooth contact pattern considering contact ratio. The region whose clearance between the pinion and gear tooth surfaces is less than 80 μm is displayed. The tooth contact patterns are obtained around the centers on the tooth surfaces of both drive and coast sides, respectively.

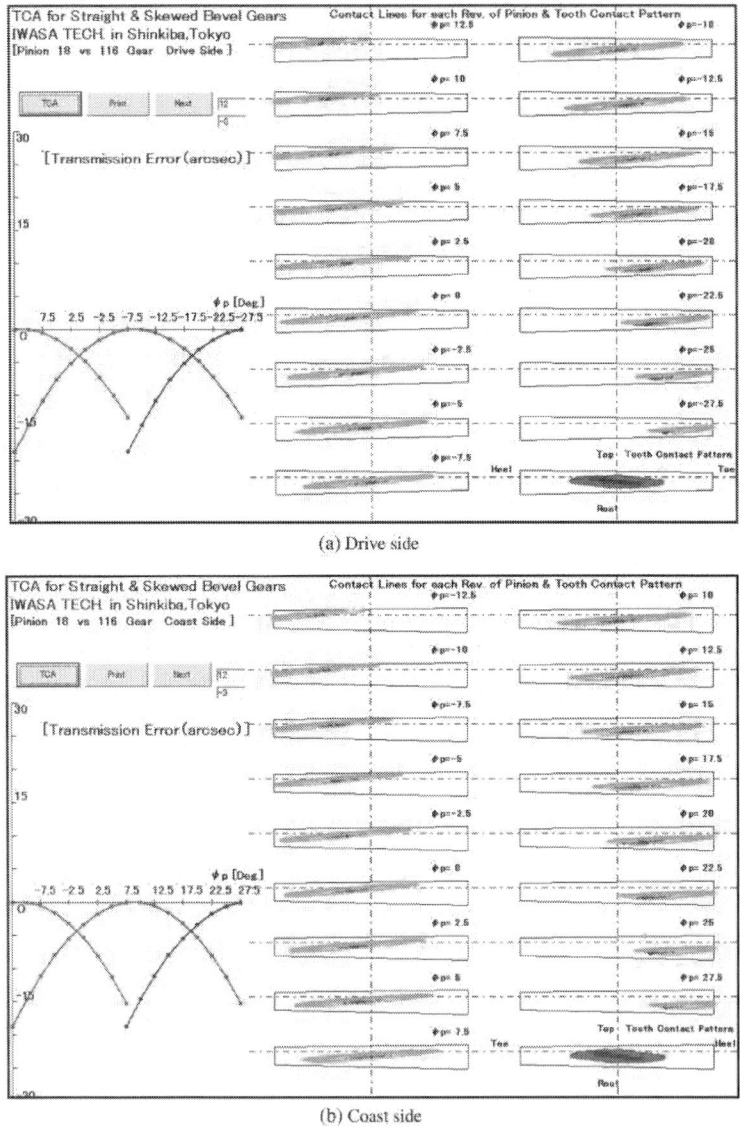

Figure 8. Analyzed results of tooth contact pattern and transmission errors without taking into account tooth surface form deviations.

The left side in Fig. 8 shows the analyzed transmission errors. The shape of transmission errors is parabolic. The parabolic transmission errors occur due to the influence of both the profile modification and the crowning. In this case, the rotation is transmitted smoothly. Therefore, it is important to have the intersection before and after meshing. The maximum value of the transmission errors is about 5 arc sec of both drive and coast sides, respectively. These transmission errors can be adjusted by changing Δc and Δs. From these results, the skew bevel gear pair has good performance in ideal conditions.

The tooth contact pattern and transmission errors were analyzed reflecting the results in Table 3. In this case, the modifications of the tooth surface and relieving were appropriately given based on the analyzed results. Table 4 shows the amounts of modification factor of pinion tooth surface. The skew angle modification $\Delta\beta$, pressure angle modification $\Delta\alpha$, amounts of tooth profile modification δc, and amounts of crowning δs were given. The amounts of relieving modification of -0.1 mm in the toe side and of -0.3 mm in the heel side were given.

Table 4. Amounts of modification factor of pinion tooth surface.

Pinion member	Skew angle [deg.] $\Delta\beta$	Pressure angle [deg.] $\Delta\alpha$	Profile mod. [mm] δc	Crowning [mm] δs	Relieving modification [mm]			
					Toe side		Heel side	
					Distance	Amount	Distance	Amount
Drive side	0.05	−0.60	0.1	−0.40	10	−0.1	80	−0.3
Coast side	0.03	−0.25	0.2	0.8	10	−0.1	80	−0.3

Fig. 9 shows the analyzed results of tooth contact pattern and transmission errors when taking into account the tooth surface form deviations in the same manner as Fig. 8. The tooth contact pattern deviates slightly from the center on the tooth surface of both drive and coast sides, respectively. However, these contact patterns seem to be acceptable in practical use because the required tooth contact pattern is satisfied comparing with that of Fig. 5. Moreover, the transmission errors become large on the coast side. These transmission errors also seem to be acceptable in practical use. Therefore, the coordinates of the tooth surface of the pinion member are defined as nominal data and the pinion member is remanufactured based on these results.

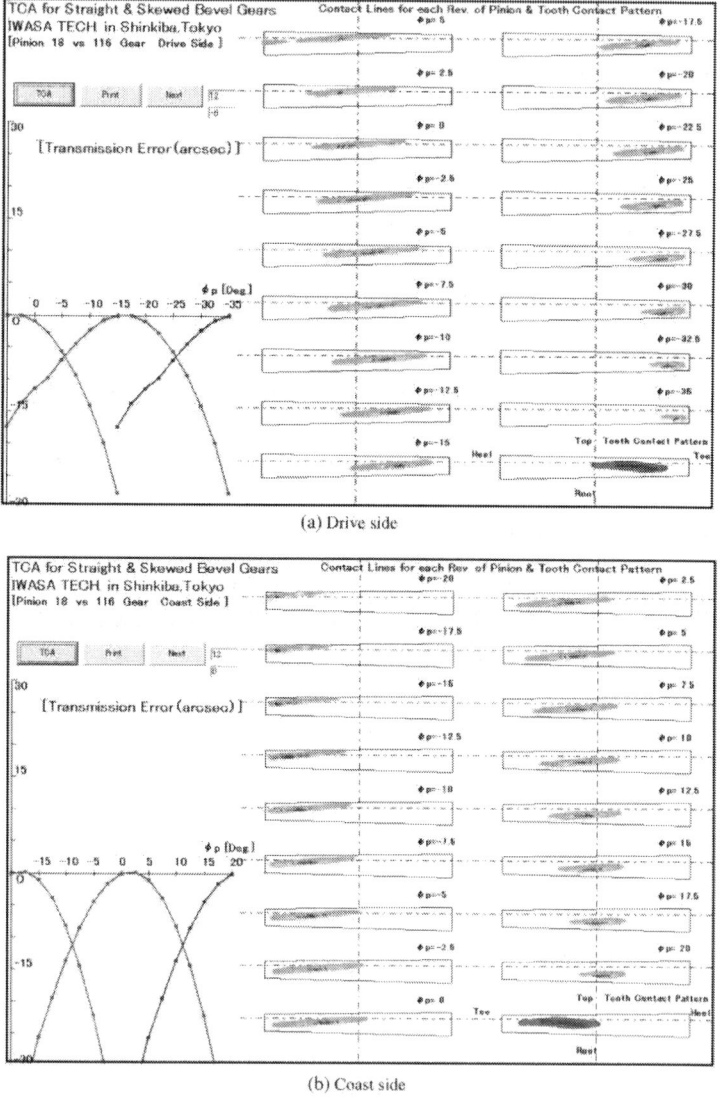

Figure 9. Analyzed results of tooth contact pattern and transmission errors with taking into account tooth surface form deviations.

REMANUFACTURING OF PINION MEMBER

The pinion member was remanufactured using a 5-axis CNC machining center (DMG Moriseki Co., Ltd. DMU210P) based on the nominal data as mentioned in Section 4.2. In this case, the reference and hole surfaces in addition to the tooth surfaces can be machined, and tool approach is provided from optimal direction using multi-axis control since the structure of the 2-axis of the inclination and rotation in addition to translational 3-axis are added. Therefore, a thicker tool can be used. This should reduce the machining time and obtain the tooth surface with less roughness. The radius end mills made of cemented carbide for a hard cutting tool were used in the machining of tooth surface. The number of edges is six, and the diameter of end mill is 10 mm. Ball end mills were used in the machining of the tooth bottom. The number of edges is six, and the diameters of end mills are 10 mm and 5 mm, respectively in the machining of the tooth bottom. The used pinion material was 18CrNiMo06. First, the pinion-work was rough-cut and heat-treated. Afterward, the pinion member was semi-finished with the axial depth of cut of 0.2 mm after heat-treatment. Finally, the pinion member was finished with the axial depth of cut of 0.05 mm by swarf cutting. Machining with high accuracy and efficiency utilizing the advantages of a CNC machining center in swarf cutting would be expected. Table 5 shows the conditions for semi-finishing and finishing of the pinion machining. Fig. 10 shows the situation of swarf cutting of the pinion member. The machining time of one side in semi-finishing was about two hours. The finishing took about six hours. The machining was finished without problems such as defects of the end mill.

Table 5. Conditions of pinion machining.

Processes	Diameter of end mill (mm)	Revolution of main spindle (rpm)	Feed (mm/min.)	Axial depth of cut (mm)	Swarf step (mm)	Time/one side (min.)
Semi-finishing	10	1400	1100	0.2	1.2	120
Finishing	10	1600	1100	0.05	0.4	360

Figure 10. Swarf cutting of pinion member.

The tooth surfaces of the remanufactured pinion member were measured using a CMM and were compared with the nominal data of the tooth surface that were determined inSection 4.2. Fig. 11 shows the measured result of the pinion member. The maximum values of the magnitude of deviations are 0.089 mm, 0.126 mm, on the drive and coast sides, respectively. These values as a whole are a sufficient permitted limit although the value on the coast side is large somewhat. The reason is that the size of the skew bevel gears is very large and seems not to have an influence on the tooth contact pattern.

REMANUFACTURING OF PINION MEMBER 45

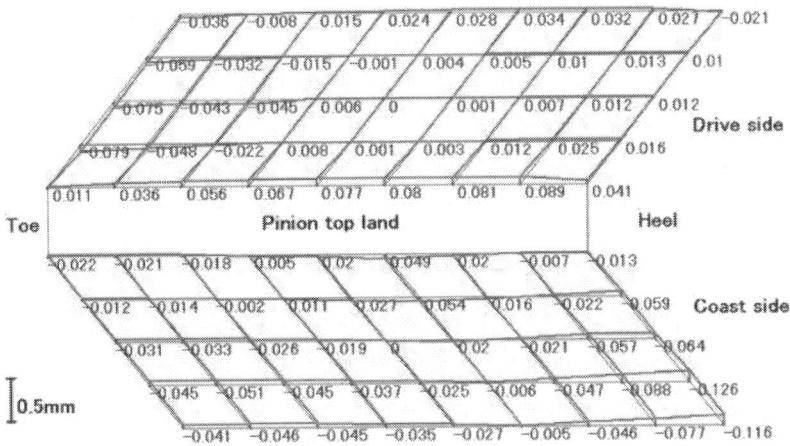

Figure 11. Measured result of pinion member.

The existing gear member and manufactured pinion were set on a gear meshing tester and the experimental tooth contact patterns were investigated. Fig. 12 shows the results of the experimental tooth contact patterns on the gear tooth surface of drive and coast sides, respectively. Although the experimental tooth contact pattern deviates from the center of the tooth surface slightly on both drive and coast sides, it is almost the same as that in Fig. 9 with respect to the tooth surface deviations. From these results, the validity of the remanufacturing method of the pinion member mating the existing gear member using a CNC machining center was confirmed.

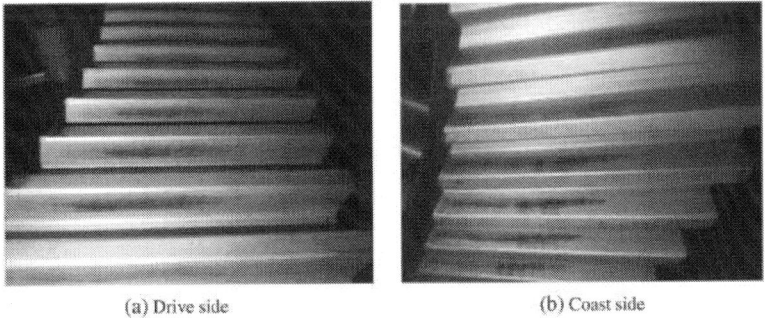

(a) Drive side   (b) Coast side

Figure 12. Experimental tooth contact patterns of existing gear and manufactured pinion.

REMANUFACTURING OF GEAR MEMBER

The gear member is also remanufactured using the nominal data of the tooth surface x_a of the existing gear member in the same manner as the pinion member in order to apply to the case where the only pinion member exists in reverse or both gear and pinion members do not exist.

First, the gear-work was rough-cut and heat-treated. Afterward, the gear member was semi-finished with the axial depth of cut of 0.2 mm after heat-treatment. Finally, the gear member was finished with the axial depth of cut of 0.05 mm by swarf cutting. Table 6 shows the conditions for semi-finishing and finishing of the gear machining. Fig. 13 shows the situation of swarf cutting of the gear member. The machining time of one side in semi-finishing was about nine hours. The finishing took about twenty six hours. The machining was finished without problems such as defects of the end mill.

Table 6. Conditions of gear machining.

Processes	Diameter of end mill (mm)	Revolution of main spindle (rpm)	Feed (mm/min.)	Depth of cut (mm)	Swarf step (mm)	Time/one side (min.)
Semi-finishing	10	1200	1100	0.2	1.8	520
Finishing	10	1400	1100	0.05	0.6	1560

Figure 13. Swarf cutting of gear member.

The tooth surfaces of the remanufactured gear member were measured using a CMM and were compared with nominal data that were determined from the real gear tooth surface formalized using polynomial expression. Fig. 14 shows the measured result of the gear member. The maximum values of the magnitude of deviations are 0.015 mm, 0.006 mm, on the drive and coast sides, respectively. These values are small and a sufficient permitted limit. The remanufactured pinion and gear members were set on a gear meshing tester and the experimental tooth contact patterns were investigated.Fig. 15 shows the results of the experimental tooth contact patterns on the gear tooth surface of drive and coast sides, respectively. The experimental tooth contact pattern showed good agreement compared with analytical one.

Machine Component Design

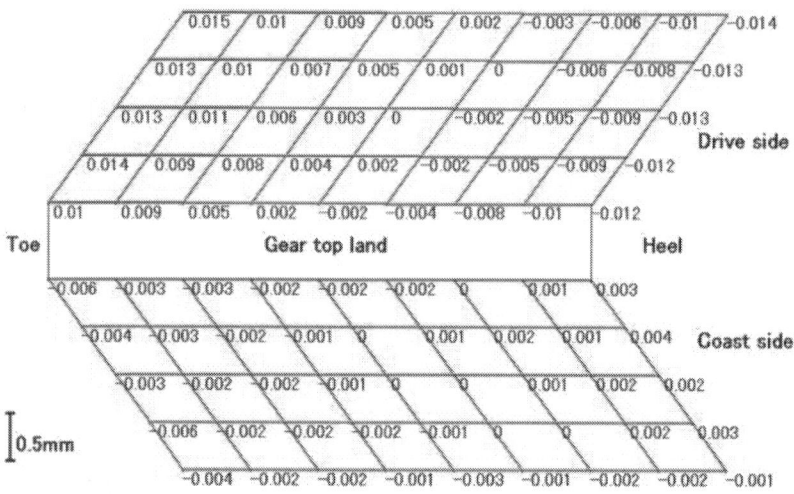

Figure 14. Measured result of gear member.

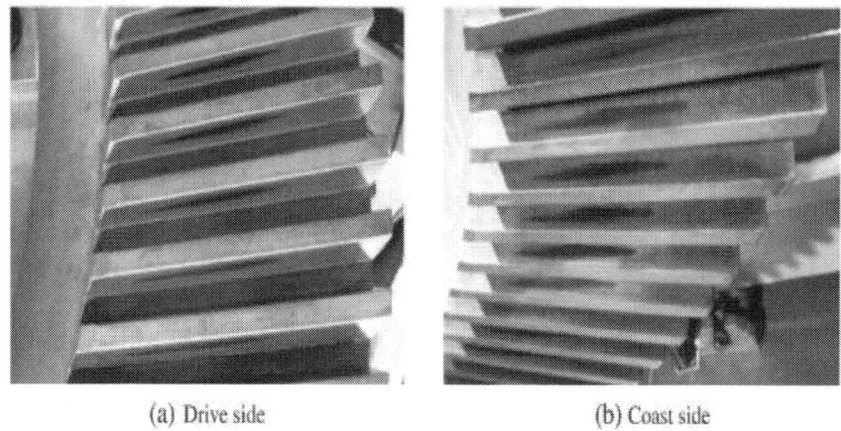

(a) Drive side   (b) Coast side

Figure 15. Experimental tooth contact patterns of remanufactured pinion and gear members.

These results are expected to produce the interchangeable skew bevel gear pair as cylindrical gears such as spur gears and helical gears.

CONCLUSIONS

In this paper, a method for remanufacturing of large-sized skew bevel gears using a CNC machining center was proposed. The main conclusions obtained are summarized as follows:

1) The tooth surface forms of skew bevel gears were modeled mathematically and simply.
2) The real tooth surface of a existing gear member was measured using a CMM and the deviations between the real and theoretical tooth surface forms were formalized as polynomial expression using the measured coordinates.
3) The tooth contact pattern and transmission errors reflecting the deviations of the tooth surface forms of the gear member were analyzed, and the tooth surface form of the pinion member that has good performance mating with the existing gear member was designed.
4) The pinion member was manufactured by swarf cutting using a CNC machining center.
5) The real tooth surfaces of the remanufactured pinion member were measured using a CMM and the tooth surface form deviations were detected. Moreover, the experimental tooth contact patterns of the existing gear member and remanufactured pinion member were compared with analytical one.
6) The gear member with the tooth surface form of the existing gear member was also remanufactured, and the deviations between the measured and nominal data were detected. Moreover, the tooth contact patterns of the remanufactured pinion and gear members were investigated.

As a result, the tooth surface form deviations of the remanufactured pinion and gear were a permitted limit on both the drive and coast sides, respectively. The experimental tooth contact patterns showed good agreement compared with analytical one.

Nomenclature

z_i number of teeth of imaginary crown gear

z_p, z_g numbers of teeth of pinion and gear

λp_0, λg_0 pitch cone angles of pinion and gear

P reference point

R_m mean cone distance

b face width

M_n normal module

α pressure angle

u, θ variable parameters representing curved line

Δc, Δs amounts of tooth profile modification and crowning

ρ_c, ρ_s radii of curvatures of circular arcs in xz and xy planes

β skew angle

X position vector of tooth surface of imaginary crown gear in O-xyz

N unit normal vector of X in O-xyz

Ψ parameter representing rotation angle of imaginary crown gear about z axis

X_ψ position vector of tooth surface of imaginary crown gear in O-$x_s y_s z_s$

N_ψ unit normal vector of X_ψ in O-$x_s y_s z_s$

W relative velocity vector between imaginary crown gear and generated gear in O-$x_s y_s z_s$

x_p position vector of pinion tooth surface

n_p unit normal vector of x_p

x_g position vector of gear tooth surface

n_g unit normal vector of x_g

$x^{(i)}$ position vector of i-th point of tooth surface in O_m-$x_m y_m z_m$

$n^{(i)}$ unit normal vector of $x^{(i)}$

$x_m^{(i)}$ position vector of i-th measured tooth surface coordinates in O_m-$x_m y_m z_m$

$\delta^{(i)}$ deviation between measured coordinates and the coordinates of theoretical tooth surface for each point on grid toward direction of normal of tooth surface

$\delta_{11}, \delta_{12}, \delta_{21}, \delta_{22}, \delta_{31}, \delta_{32}, \delta_{41}$ parameters defining deviation

$a_{11}, a_{12}, a_{21}, a_{22}, a_{31}, a_{32}$ fundamental components of polynomial expression

X, Y variable parameters in direction of tooth trace and tooth profile

H, T ranges of evaluation of tooth surface in X and Y directions

b_1, b_2, b_3 fundamental components of polynomial expression

$A_1, A_2, A_3, A_4, A_5, A_6$ fundamental components of polynomial expression

x_a position vector of theoretical tooth surface in consideration of tooth flank form errors

ϕ_p rotation angle of pinion about y_h axis in $o_h\text{-}x_h y_h z_h$

ϕ_g rotation angle of gear about z_h axis in $O_h\text{-}x_h y_h z_h$

$\Delta\phi_g$ function of transmission error

B coordinate transformation matrix for rotation about y_h axis

C coordinate transformation matrix for rotation about z_h axis

$\Delta\beta$ amount of modification of β

$\Delta\alpha$ amount of modification of α

δc amount of modification of Δc

δs amount of modification of Δs

REFERENCES

1. Y.C. Tsai, P.C. Chin, Surface geometry of straight and spiral bevel gears, Trans. ASME J. Mech. Transm. Autom. Des. 109 (1987) 443–449.
2. D.P. Townsend, Dudley's Gear Handbook, The Design, Manufacture, and Application of Gears2nd Ed, McGraw-Hill, New York, 1991. 2. 9–2. 17.
3. Stephen P. Radzeevich, Handbook of Practical Gear Design and Manufacture, 2nd Ed CRC Press, 1994. 53.

4. Andrew D. Dimarogonas, Machine Design A CAD Approach, Wiley-InterScience, 2001. 869–874.
5. J.R. Davis, Gear materials, properties, and manufacture, ASM International Technical Books Committee, USA 2005, pp. 92–99.
6. A. Fuentes, J.L. Iserte, I. Gonzales-Perez, F.T. Sanchez-Marin, Computerized design of advanced straight and skew bevel gears produced by precision forging, Comput. Meth
7.
8. ods Appl. Mech. Eng. 200 (2011) 2363–2377.
9. M. Nakaminami, T. Tokuma, T. Moriwaki, K. Nakamoto, Optimal structure design methodology for compound multiaxis machine tools — I (analysis of requirements and specifications), Int. J. Autom. Technol. 1 (2) (2007) 78–86.
10. T. Moriwaki, Multi-functional machine tool, CIRP Ann. 57 (2) (2008) 736–749.
11. S.H. Suh, W.S. Jih, H.D. Hong, D.H. Chung, Sculptured surface machining of spiral bevel gears with CNC milling, Int. J. Mach. Tools Manuf. 41 (6) (2001) 833–850.
12. J.T. Alves, M. Guingand, J. Vaujany, Designing and manufacturing spiral bevel gears using 5-axis computer numerical control (CNC) milling machines, J. Mech. Des. Trans. ASME 135 (2) (2013) 024502-1-6.
13. B. Lei, G. Cheng, H. Lowe, X. Wang, Remanufacturing the pinion: an application of a new design method for spiral bevel gears, Adv. Mech. Eng. 2014 (2014) 257581-1-9.
14. K. Kawasaki, I. Tsuji, Y. Abe, H. Gunbara, Manufacturing method of large-sized spiral bevel gears in cyclo-palloid system using multi-axis control and multi-tasking machine tool, Proc. of International Conference on Gears, Garching, Germany, 1 2010, pp. 337–348.
15. F.L. Litvin, C. Kuan, J.C. Wang, R.F. Handschuh, J. Masseth, N. Maruyama, Minimization of deviations of gear real tooth surfaces determined by coordinate measurements, J. Mech. Des. Trans. ASME 115 (4) (1993) 995–1001.
16. C.-Y. Lin, C.-B. Tsay, Z.-H. Fong, Computed-aided manufacturing of spiral bevel and hypoid gears by applying optimization techniques, J. Mater. Process. Technol. 114 (1) (2001) 22–35.
17. Q. Fan, Tooth surface error correction for face-hobbed hypoid gears, J. Mech. Des. Trans. ASME 132 (1) (2010) 011004-1-8.
18. V. Simon, Influence of tooth modification on tooth contact in face-hobbed spiral bevel gears, Mech. Mach. Theory 46 (2011) 1980–1998.
19. A. Artoni, M. Gabiccini, M. Kolivand, Ease-off based compensation of tooth surface deviations for spiral bevel and hypoid gears: only the pinion needs corrections, Mech. Mach. Theory 61 (2013) 84–101.

20. I. Tsuji, K. Kawasaki, H. Gunbara, Manufacturing method of pinion member of large-sized skew bevel gears using multi-axis control and multi-tasking machine tool, Proc. of the AGMA Fall Technical Meeting, Dearborn, USA, CD-ROM 2012, p. 15.
21. I. Tsuji, K. Kawasaki, H. Gunbara, H. Houjoh, S. Matsumura, Tooth contact analysis and manufacture on multitasking machine of large-sized straight bevel gears with equi-depth teeth, J. Mech. Des. Trans. ASME 135 (3) (2013) 034504-1-8.
22. T. Sakai, A study on the tooth profile of hypoid gears, Trans. JSME 21 (102) (1955) 164–170 (in Japanese).
23. F.L. Litvin, A. Fuentes, Gear Geometry and Applied Theory, 2nd Ed. Cambridge University Press, UK, 2004. 98–101.
24. K. Kawasaki, I. Tsuji, Analytical and experimental tooth contact pattern of large-sized spiral bevel gears in cyclo-palloid system, J. Mech. Des. Trans. ASME 132 (2010) 041004-1-8.
25. Q. Fan, S. DaFoe Ronald, W. Swanger John, Higher-order tooth flank form error correction for face-milled spiral bevel and hypoid gears, J. Mech. Des. Trans. ASME 130 (2008) 072601-1-7.
26. H.J. Stadtfeld, Handbook of Bevel and Hypoid Gears, Calculation, Manufacturing and Optimization, Rochester Institute of Technology, R·I·T, 1993. 9–12.
27. A. Artoni, A. Bracci, M. Gabiccini, M. Guiggiani, Optimization of the loaded contact pattern in hypoid gears by automatic topography modification, J. Mech. Des. Trans. ASME 131 (2009) 011008-1-9.

CITATION

Kazumasa Kawasaki, Isamu Tsuji, Hiroshi Gunbara, Haruo Houjoh, Method for remanufacturing large-sized skew bevel gears using CNC machining center, Mechanism and Machine Theory, Volume 92, October 2015, Pages 213-229, ISSN 0094-114X, http://dx.doi.org/10.1016/j.mechmachtheory.2015.05.013.

CHAPTER 3

Circular Causality and Indeterminism in Machines for Design

Thomas Fischer

Department of Architecture, Xi'an Jiaotong-Liverpool University, Suzhou 215123, China

ABSTRACT

Presenting a hard-to-predict typography-varying system predicated on Nazi-era cryptography, the Enigma cipher machine, this paper illustrates conditions under which unrepeatable phenomena can arise, even from straight-forward mechanisms. Such conditions arise where systems are observed from outside of boundaries that arise through their observation, and where such systems refer to themselves in a circular fashion. It argues that the Enigma cipher machine is isomorphous with Heinz von Foersters portrayals of non-triviality in his non-trivial machine (NTM), but not with surprising human behaviour, and it demonstrates that the NTM does not account for spontaneity as it is observed in humans in general.

BACKGROUND

From the inside, it can be challenging to determine the scope, shape and development of the field one is operating in. Are the frontiers of architectural design research static, unvarying limits? Or are the frontiers of architectural design research changing borderlines, shifting according to modes, depths and directions of enquiry? To what extent do its design and research aspects overlap, and to what extent are design and research comparable or compatible? Do design and research have enough in common to be approached as equals, rendering insights into one of them applicable to the respective other? Are they viable models or metaphors for one another, or are they too different to allow such analogies between them? Answers to these questions, of course, depend much on what is meant by design and by research.

Understandings of design and research, of their methods, tools and standards, diverge considerably in different contexts. The argument presented here addresses design, design tools and research methods in reference to systemic boundaries and circular re-entry, and with regards to the notion of determinability in order to shine a critical light on those instances where design and design research are approached in terms of purely linear cause and effect. It is shown that conceptualisations of design (research) in terms of (natural-scientific or computational) linear causality may be unduly limited.

The argument below draws parallels between the designing human mind and a mechanical (cipher) machine. This is not to say that the mind is like a mechanism, or that mechanisms can act in the ways human minds do. The point made is merely that minds and some mechanisms are characterised by circular re-entry, which, in both cases, leads to indeterminable behaviour, i.e. novelty. Neither circular causality nor indeterminism, however, is recognised by natural-scientific reasoning.

SYSTEM BOUNDARIES, INPUT, OUTPUT AND RE-ENTRY

Computer-aided architectural designing is an endeavour in which the boundaries of systems are crossed. "System" is understood here as whatever set of elements an observer considers to act together, following a common goal. An observer may choose to regard the components that make up a computer as a system. Similarly, an observer may choose to regard the organs making up the organism of a designer as a system, or consider the designer and the computer together as a system. With these different ways of looking (Weinberg, 2001), the imaginary boundary that circumscribes what is regarded as a system changes, and what is considered as a system lies in the eyes of the observer. Sometimes there are physical boundaries containing what is regarded as a system, such as the skin of a designer and the case of a computer but this is coincidental. Designer and computer together may be regarded as one system contained by an imaginary, but without a physical boundary. Patterns in the widest sense crossing the imaginary boundaries of systems are, depending on perceived direction, called inputs and outputs.

A common example of systems whose boundaries are crossed by incoming inputs and by outgoing outputs is the behaviourist-type stimulus-response structure of the kind shown on the left-hand sides of Figure 1 and Figure 2. This structure offers convenience in modelling various systemic relationships not only by way of abstraction and of being broadly applicable. It is also conveniently compatible with common basic tools of rational modern thought such as linear logic and syllogistic reasoning. Humans are frequently described as systems which, prompted by input, produce output. And, typically, so are computers. Alternatively, although this happens less frequently, an observer may also choose to view multiple systems (inter)acting together as one system which responds to input by producing output. Human–computer interaction in CAAD may be viewed in this way, along the lines of the following statement by Bateson (1972, p. 317): "The computer is only an arc of a larger circuit which always includes a man and an environment from which information is received and upon which efferent messages from the computer have effect." Other examples in the design context include the interactions between members of a design team, and the interaction between a designer and his or her sketching (Fischer, 2010, p. 612).

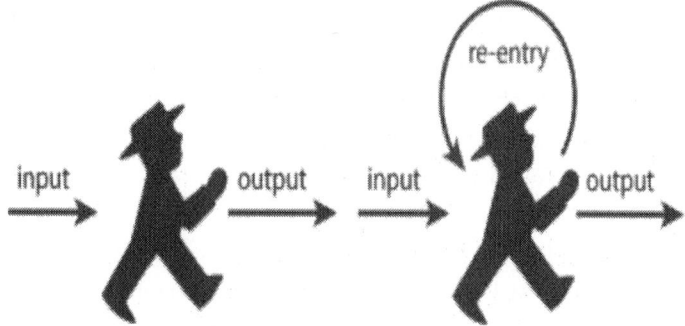

Figure 1. Human viewed as a linear stimulus–response system (left) and with the acknowledgement of circular self-reference (right).

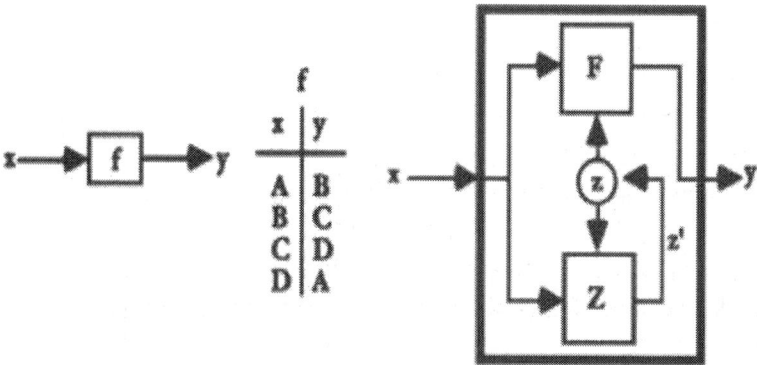

Figure 2. Trivial machine with truth table (left) and non-trivial machine (right), reproduced from Von Foerster (2003, pp. 310–311).

Any of these systems – human, computer, human–human, human–computer and so on – is defined by an imaginary boundary projected by an observer. This imaginary boundary sets the system's interior apart from its exterior. If a human considers herself or himself as a system, then making (the interior self-affecting the exterior other) and learning (the exterior other affecting the interior self) constitute instances of outputs and inputs crossing boundaries. While cyclical relationships such as the ones observable in human–computer interaction are commonly dissected and broken up into pieces, it is uncommon to turn systems back on themselves to form closed loops. This is because modern culture appreciates systems which allow description in terms of linearly-causal logic and which offer predictable control in terms of defined states. Closed loop structures tend to be appreciated only where they facilitate control, typically in the form of negative feedback and error correction or of stable oscillation. Unpredictable fluctuations and out-of-control patterns tend to be unwelcome outside of artistic and experimental domains. They are rarely the subject of formal analysis, and attempts at their formal analysis are hampered by the linear nature of common tools of description. Nonetheless, the (designing) human mind must be acknowledged not merely as a static stimulus-response system, as a static translator between inputs and outputs, but as a system whose input channels are subjected to its own output. Contrary to the technologies it currently tends to develop, the human mind is subjected to what it itself produces and is thus changed by its own performance (see Figure 1).

As stated above, design, being at least in part out-of-control (Glanville, 2000), involves not only linear but also circular causality – between design team members, between designers and their sketches etc. (Fischer, 2010). Common algorithmic devices for generative, computer-based design likewise involve circular feedback such as the potentially circularly-causal relationship between any two cells in a cellular automata system, or the self-referential relationships in L-systems, in evolutionary algorithms and so on. Input–output operations can leave traces inside of (designing) systems equipped with suitable "internal state" memory. Such systems can therefore, in effect, become different systems through each of their operations. And through the interaction between input and/or output with given internal states, such machines can behave unpredictably. Systems of this kind will be explored and illustrated in the following, with special attention to the limits of purely mechanical or digital implementations in the design context.

The loop which is formed when human articulations feedback as an input to the human creative process allows expressions of the mind to re-enter into the mind where they may leave increasingly stable traces (Glanville, 1997, p. 2), i.e. memory. This view was substantiated in Von Foerster (1950)'s interpretation of a previous study of human memory. In that previous study subjects had been asked to memorise random, meaningless syllables and to re-count as many of them as they could afterwards at regular intervals. Memory and progressive forgetting were shown to follow an exponential decay curve, which did not approach zero, but a number of syllabi greater than zero that the subjects were increasingly more likely to remember permanently. Von Foerster explains this with the human being capable of both input (listening) and output (speaking), and hence circular closure and re-entry of articulations. Thus, every re-counting of a remembered syllable (output) is also a new input which reinforces what is known. Repeated recalling thus leads to an eventually stable subset of remembered syllables.

Von Foerster (2003, p. 311) illustrated processes of this nature using his notion of the trivial machine (TM), which he juxtaposed to his notion of the non-trivial machine (NTM). Somewhat comparable to Turing's (1937, p. 231ff.) proposal of the Turing Machine, von Foerster describes both the TM and the NTM as minimal hypothetical machines not for the purpose of implementation, but for the purpose of illustrating ideas. He describes both TM and NTM as basic input–output (stimulus-response) systems, each being a mechanism connected to an input channel and an output channel. The TM predictably translates inputs into corresponding outputs, so that an external observer can, after a period of observation, establish clear causal relationships between possible inputs and resulting outputs, for example in the form of a "truth table" as shown on the left of Figure 2. A complete truth table is a reliable model for predicting the TM's output responses to given inputs, irrespectively of how long the machine has been in operation. In contrast, the NTM contains means to memorise a machine state (labelled z on the right of Figure 2). This state co-determines the machine's output together with its input. At the same time, the state may change with each input–output operation. This results in a vast number of possible input–output mappings which can easily exceed the quantitative limits of what an external observer can determine analytically, i.e. derive predictive capabilities from Glanville (2003, p. 99). The NTM's history of input–output translations can be said to leave traces in the machine, which in effect turns into a different machine through and for each of its own operations. An outside observer cannot easily establish a reliable truth table by which outputs resulting from given inputs can be predicted.

Von Foerster's presentations of the NTM changed slightly from presentation to presentation, in particular with regards to that which brings about state changes in the machine. According to Von Foerster (1970, p. 139) the transitions of internal states depend on the machine's previous state and on its input, according to Von Foerster (1972, p. 6) internal state transitions depend on the machine's previous output, and according to Von Foerster (1984, p. 10) internal state transitions depend again on the machine's previous state and on its input.

Another description of non-triviality offered by von Foerster is the image of a schoolboy who displays non-trivial behaviour by responding to a maths or history problem with an unexpected answer. Similar to the mechanistic portrayal of non-triviality, the anthropomorphic portrayal changes slightly from one presentation to the next. In Sander (1999) and Pruckner (2002) the schoolboy responds to the problem "2 times 2" with the result "green". In Von Foerster (2003, p. 311) he responds to problem "2 times 3" with the answer "green" or with the answer "Thats how old I am". In Von Foerster (1972, p. 6) he responds to the question When was Napoleon born? with the answer "Seven years before the Declaration of Independence." Von Foerster deplores the state of educational systems in which school children who offer such unexpected responses are deemed insufficiently trivialised, and therefore trained more until they produce the desired answers reliably.

The distinction between the trivial and the non-trivial behaviour deserves attention in general and in particular in our field because in every encounter the choice between both metaphors determines much of the ethical stance one takes towards others, tools, buildings etc. Virtually all of our science and technology corresponds to the principle of the trivial machine in the sense that a given input is expected to always reliably lead to the same output. Multiple computers are, put simply, expected to always have the same response in the face of the same task or problem. Stereotypical engineers, managers and representatives of other professions, be they allied with architecture or not, are likewise expected to arrive at the same results when presented with the same input. In the education of these professions this aspiration to the ideal of the trivial machine is enforced with the principle of scientific repeatability. Multiple stereotypical engineers tasked with the same structural analysis problem, or the same engineer tasked with the same structural analysis problem twice should reliably arrive at the same results, i.e. fulfil expectations predictably and reliably. Briefing multiple architectural designers with the same project brief, in contrast, makes sense only if variety among multiple responses is desired. Briefing the same architectural designer with the same task twice will also lead to two different outcomes because the second time around, one would be facing a different architect one who was subjected to her/his own first design process and outcome, which left traces in her/him and thus changed her/him. In this sense, and in the sense of Heraclitus statement that "No man ever steps in the same river twice, for it's not the same river and he's not the same man", one cannot design (or learn, for that matter) the same thing twice.

Von Foerster explains convincingly that neither NTM nor schoolboy permits analytical determination from the perspective of a human observer. He does not, however, address the possible conclusion that NTM and unpredictable human are therefore to be taken as isomorphic. He calls for humans to be perceived as non-trivial (Von Foerster, 1972, p. 6), without addressing the question of whether the NTM would be capable of giving the answers given by the schoolboy.

VARIETY AND AUTOMATED PRODUCTION

Surprising variety (in the cybernetic sense: number of choices available) and reliable predictability are, paradoxically both for better and for worse, essential human needs and human characteristics (Fischer, 2010, p. 611). We experience this paradox in numerous contexts in which we enjoy both stimulating variety in expression as well as economic and organisational benefits of uniformity. In shaping our products and environments, the advantages offered by predictably uniform (hence interchangeable) prefabricated components famously gave rise to assembly-line based production since the early days of industrial production; and it is part and parcel of architectural construction today. Having been introduced to architecture with uniform building elements, prefabrication brought along with it sameness at the scales of component repetition. At small scales of component repetition, such as that of clay bricks (Fischer, 2007), interchangeability may be appreciated for allowing flexibility and subtle texture. At larger scales such as that of floor plans or whole buildings as found in Platenbau developments (Hopf and Meier, 2011), repetitive sameness is criticised for being monotonously boring or even socially detrimental. Aiming at repeatable input–output relationships, early computer applications in our field focussed on predictable input–output relationships, leading to criticisms of applying the computer as a "fancy drawing board" (Dantas, 2010, p. 161) and of valuing it as an equivalent to "an army of clerks" (Alexander, 1965). In architecture and in other industries, there are now tendencies acting against monotonous sameness. Referred to as customisation approaches (Gilmore et al., 1997), these tendencies are increasingly aided by computational (generative, parametric etc.) techniques that allow increasing of variety via circular feedback. The development of typography follows a similar pattern. Moveable type introduced economic benefits along with monotonous sameness to book printing. Using type wheels and the like, typewriters, teletypes and computer printers achieved similar predictable sameness and cost-efficiency also in documents produced in small numbers. Contextual variations such as ligatures have been introduced to mechanical typesetting. Some contemporary computer typefaces go further and achieve "organic", "random" or "handwritten" appearances by introducing randomness to curve paths or by providing sets of alternative glyphs for the same characters.

ENIGMA CIPHER MACHINE

With these working principles, the NTM is essentially isomorphous with the Enigma machine (Scherbius, 1928; Fischer, 2012) which was used to encipher and to decipher communications in Nazi Germany before and during World War II (this relationship between NTM and Enigma machine was previously suggested by Tessmann, 2008, p. 55). Looking somewhat like a typewriter, the Enigma machine was used to encrypt and to decrypt text messages by substituting letters with a replacement mechanism that changes systematically as the machine is used. It takes its input via a qwertz keyboard (label 1 in Figure 3) with typically 26 keys, and offers its output via typically 26 lamps which are also arranged in a qwertz layout (label 12 in Figure 3). Pressing any key closes an electrical circuit which travels across a set of cylindrical rotors each of which contains a different irregularly-connected wiring, leading to the illumination of a lamp with a different letter. Before it closes a circuit each keystroke also results in the change of the internal state of the machine by way of rotating one or more of the rotors by one twenty-sixth of a full rotation so that the combined irregular wiring changes for each letter that is enciphered or deciphered. Additionally, a plug board allows the swapping of pairs of letters using patch cables. Much like von Foerster's NTM, the Enigma machine translates input characters to output characters, with every translation resulting in a re-mapping of the set of accepted input characters to the set of available output characters. The Enigma machine demonstrates that the NTM is implementable as a physical device, which is very challenging to determine analytically from the perspective of an external observer.

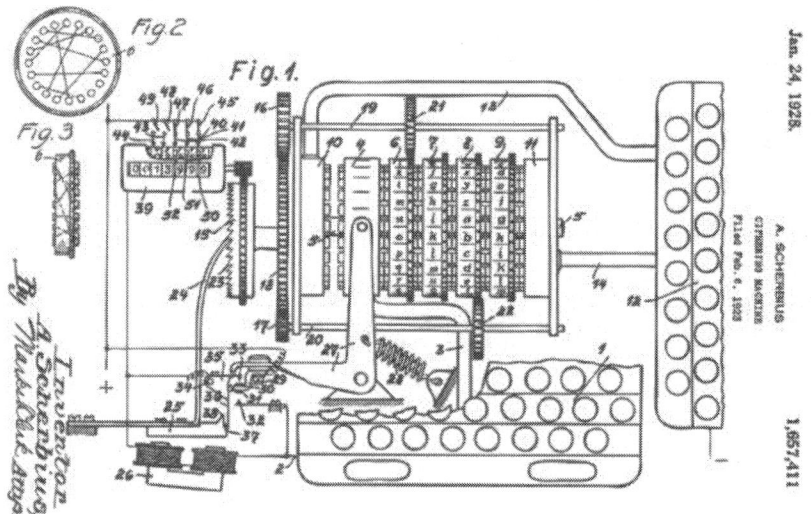

Figure 3. Schematic diagram of Enigma machine (from Scherbius, 1928).

Pressing a key will activate one of the lamps, apparently at random, according to selection of cylinders and their current orientation. Additionally, each keystroke results in the rotation of the first cylinder by one of 26 rotation positions, after 26 keystrokes, the second cylinder will also rotate by one position and so forth, somewhat in the fashion of the digit cylinders in a mechanical odometer. Thus, each keystroke results in a new wiring between keyboard and lamps coming into effect for the subsequent keystroke. In other words: use of the machine leaves a trace in it, changing the wiring of the machine, and hence the cipher, progressively. (Due to the symmetrical setup of the wiring going into the cylinders and back out through the same cylinders, the same machine setup can be used both to cipher and to decipher. The identical setup is achieved by referring to a secret timetable based code book of which both ends must hold a copy.) To an outside observer the input-to-output mapping of the Enigma machine is extremely difficult to determine, while it is perfectly determinable to those who developed it and who have a good understanding of its setup and inner workings. With inner workings of this kind the Enigma machine shares key characteristics of designing, making it a useful metaphor for the purpose of showing how designing is a relatively straight-forward process when viewed from the inside perspective but mysterious and wonderful when viewed from the outside perspective.

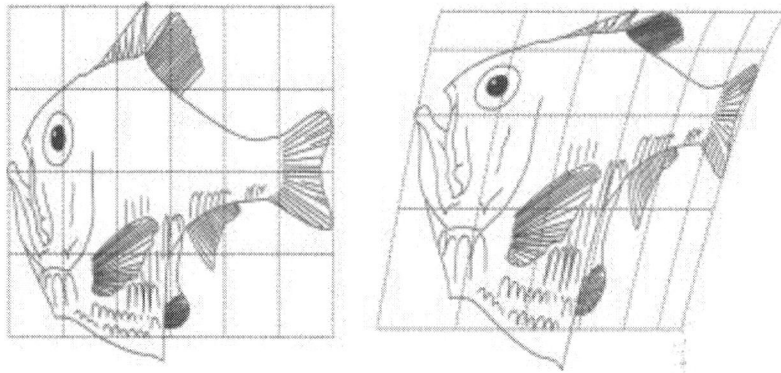

Figure 4. Fish transformations based on Thompson (1992, pp. 1053–1093).

A TYPOGRAPHICAL METAPHOR

The illustration presented here is a piece of software predicated on the Enigma machine and implemented as a VBA script controlling Rhino3D. Glyph renderings of characters input via keyboard are distorted dynamically and individually, with the use of a "private key" string stored inside the system. Somewhat akin to the (de)ciphering process of the Enigma machine, each key that is typed, modelled, transformed and rendered changes the internal state of the system (leaves a trace in it) to change the way the following glyph is distorted. In contrast to common computer typefaces, glyphs of same characters are unpredictably variant. As a point of departure, the system uses the typeface Helvetica to derive initial glyph outline curves for each typed character. The system then applies a combination of six (Thompson, 1992) transformations (see Figure 5) to these outline curves.

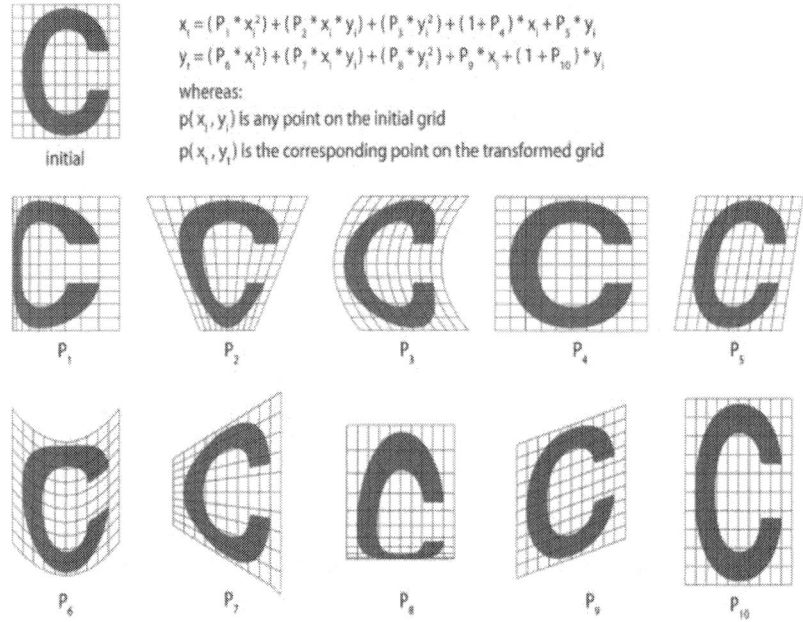

Figure 5. Quadratic functions and Thompson transformations based on parametric variation.

Thompson transformations (parametric "warping" based on quadratic functions,Wilkinson, 2005, pp. 223–224) as illustrated in Figure 3 allow positive and negative transformations based on parameters P1–P10. Of these, the presented generative system uses the six parameters P1, P2, P3, P6, P7 and P8. Parameters P4, P9 and P10 are ignored while P5 can be toggled manually, providing an "italics" option. Parametric input for the six Thompson transformations performed by the system is derived from the ASCII bit patterns of characters of a "private key" string, which is "rotated" by three characters with each input keystroke. Any ASCII string of any length can be used for this purpose (Figure 6).

A TYPOGRAPHICAL METAPHOR

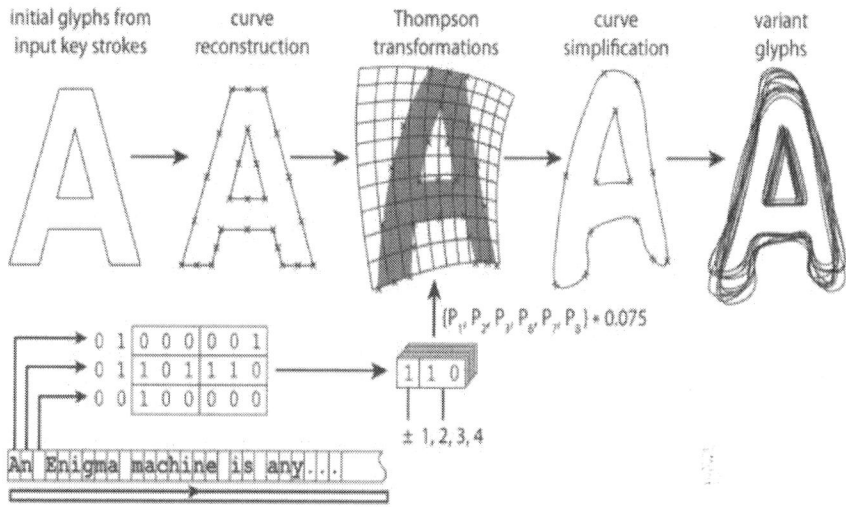

Figure 6. Key operations of glyph-variation based on "private key" string.

Once a key is pressed, the first three characters of the "private key" string are converted to their respective ASCII bit patterns and the six last bits of each are used to produce two factors out of the set −4, −3, −2, −1, 1, 2, 3, 4, which are multiplied with a scaling factor (0.075 gives a good effect) to produce a total of six parameters (see Figure 4).

Following parametric Thompson transformations, the resulting glyph outline curves are simplified by reducing their numbers of control points, resulting in casual looking, "blobby" glyphs. These are variant with identical characters being shown as different glyphs (see the bottom line of Figure 7). Overall, the resulting types are nevertheless largely consistent and recognisable as members of the same typographical style, which I call Polymorph. The generative system includes a rudimentary automatic kerning function, the performance of which is also visible in the bottom line of Figure 7. The top line of Figure 5 shows the typeface Helvetica. The middle line shows the hand-drawn, Helvetica-inspired typeface YWFT HLLVTKA Round, each character of which looks irregular, while identical characters are rendered with the same glyphs.

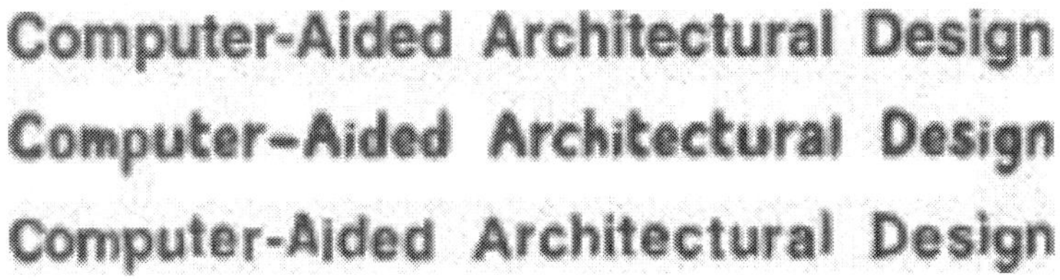

Computer-Aided Architectural Design
Computer-Aided Architectural Design
Computer-Aided Architectural Design

Figure 7. Helvetica (top), YWFT HLLVTKA round (middle) and polymorph (bottom).

Just as the Enigma machine's cipher output is enigmatic and unpredictable to outsiders such as wartime enemies, the glyph transformations generated by the described system are unlikely to be predictable to outsiders of the system. Nevertheless, both systems are perfectly determinable and appear straight-forward to those aware of both systems' setups and of the ways in which the performances of both systems leave traces within them, changing their internal states, thus in effect leading to new inner workings with each operation. Designers, articulating and (re-)considering ideas can be seen as embodying a similar, non-trivial re-entry structure which, similarly, can appear either surprising and unpredictable or straight-forward and traceable depending on an observer's inside or outside perspective.

OBSERVATIONS

Some processes are linear, predictable and seem causal while others involve circularity appear unpredictable. The difference can be shown with the distinction between the trivial machine and the non-trivial machine. Like the Enigma machine, instances of designing can be viewed as circular systems (Glanville, 1992, Fischer, 2010 and Gänshirt, 2011) which display the structure and quality of the non-trivial machine. It was demonstrated here that circular re-entry affecting the internal state of a system is a sufficient condition of indeterminability, and for systems characterised by circular re-entry to transcend the narrow notion of linear cause and effect typically applied to (digital) technology and natural-scientific research. The wonder and surprise offered by indeterminable systems depend on their interior workings evading observation (Fischer, 2008). Design is, on the inside, concerned with what is unpredictable to outsiders. It hence corresponds to the non-trivial machine. Science, always on the outside, is concerned with prediction and hence corresponds to the trivial machine. This poses a challenge to scientific researchers aiming to research into designing objectively, somewhat comparable to the challenge of cryptography that leaves outsiders mystified while insiders understand. Design processes can be appreciated and understood on the subjective inside. Objective scientific description, though, is required to approach design from the outside.

Von Foerster's TM, his NTM, and the Enigma machine share the trait that, the question of predictability notwithstanding, the varieties of acceptable inputs, of available internal states, and of their sets of possible outputs is predefined by the makeup of the machine in question. These varieties are finite and do not change through the machines operations. The Enigma machine, for example, can neither be expected to cope with unforeseen inputs that are not supported in its set of acceptable input characters, nor to spontaneously transcend its set of available outputs to include for example Chinese characters, let alone previously unknown characters. Similarly, neither the NTM, nor the TM, can be expected to accept and to process inputs other than those these machines were set up to accept and to process or to offer outputs other than those they were designed to offer.

The human mind is different in this respect (and obviously in other respects, too[1]). It has the capability of accepting previously not accepted inputs, and of expressions beyond the range of expected outputs as illustrated by the statement "2 × 2=grün". More overtly, nonetheless evident in human learning, the mind also modulates its repertoire of internal attitudes towards inputs it encounters, i.e. its range of internal states. In other words, our nervous system has the capability of amplifying the variety of ranges of inputs it accepts, of its internal states, and of outputs it can be expected to offer. The TM's, the NTM's and the Enigma machine's clearly specified, constant input, internal state, and output varieties are typical of digital technology (Fischer, 2011), which is developed and used precisely for the predictable control it offers, at the expense, as Glanville (2009, p. 119)argues, of variety. Fixes varieties in technical systems are established by a kind of observer (matchmaker) who intentionally brings system together to serve purposes by way of control (Fischer, 2011). The human mind, in contrast, can actively modulate these varieties. It can, for instance, refuse to answer a yes-or-no question in those terms or answer an arithmetic problem by naming a colour. Giving a human an arithmetic problem to solve implicitly aims to reduce that humans input and output variety to the language of arithmetic and numbers within which a response may then be evaluated. A humans concession to answer in terms of mathematics and numbers constitutes a reduction of that humans output variety. A surprising (i.e. substantially or formally incorrect) answer will then likely be dismissed as wrong, regardless of whether the contemplations that led to it have value. On such grounds Dostoyevsky's (2009, p. 25) underground man can be dismissed when he states: "I admit that twice two makes four is an excellent thing, but if we are to give everything its due, twice two makes five is sometimes a very charming thing too."

Drawing a mutually-exclusive distinction between the trivial and the non-trivial, however, von Foerster places both his mechanistic and his anthropomorphic portrayal of non-triviality in the same category, suggesting that humans are a subset of, and hence isomorphous with, non-trivial machines. While the human nervous system and non-trivial mechanisms share some characteristics, they are set apart by others, rendering them not isomorphous. Mechanistic systems such as the Enigma machine, most critically, do not share the human capability to reduce and to amplify the variety of ranges of accepted inputs, of internal states, and of expectable outputs (to make new and to drop old distinctions). As a model for human inventiveness von Foersters mechanistic description of the NTM is therefore crude at best.

While at a theoretical level the argument presented here shines a critical light on those instances where design and design research are approached in terms of purely linear cause and effect, it also offers a path forward for future research into digital design tools and computational creativity: reduction and amplification of input and output channel variety are well within the scope of technical implementability (consider digital sound or image input and output based on different sampling rates and resolutions). As a next step in the presented line of enquiry, the potential of changing input and output channel variety in systems allowing for circular re-entry will be investigated with regards to the potential for novelty generation.

REFERENCES

1. Alexander, C., 1965. The question of computers in design. Landscape 14 (3), 6–8.
2. Bateson, G., 1972. Steps to an Ecology of Mind. University of Chicago Press, Chicago.
3. Dantas, J.R., 2010. The end of Euclidean geometry or its alternate uses in computer design. In: Proceedings of SIGraDi. Bogot, Colombia, pp. 161–164.
4. Dostoyevsky, F., 2009. Notes from the Underground. Hacket, Indianapolis. Fischer, T., 2007. Rationalising bubble trusses for batch production. Autom. Construc. 16, 45–53.
5. Fischer, T., 2008. Obstructed Magic. In: Nakpan, W. (Ed.), Proceedings of CAADRIA2008. Pimniyom Press, Chiang Mai, pp. 609–618.
6. Fischer, T., 2010. The interdependence of linear and circular causality in CAAD research: a unified model. In: Dave, B. (Ed.), Proceedings of CAADRIA2010. CUHK, pp. 609–618.
7. Fischer, T., 2011. When is analog? When is digital? Kybernetes 40 (7,8), 1004–1014.
8. Fischer, T., 2012. Design Enigma. A typographical metaphor for epistemological processes, including designing, In: Fischer T. et al., (Eds.), Proceedings of CAADRIA 2012, Hindustan University, Chennai, 679–688.
9. Gänshirt, C., 2011. Werkzeuge für Ideen. Birkhäuser, Basel. Gilmore, H.G., Pine II, B.J., 1997. The four faces of mass customization. Harvard Bus. Rev. (January–February), 91–101.
10. Glanville, R., 1992. CAD abusing computing. In: Martens, B. (Ed.), CAAD Instruction: The New Teaching of an Architect? eCAADe Proceedings. Barcelona, pp. 213–224.
11. Glanville, R., 1997. A ship without a rudder. In: Glanville, Ranulph, de Zeeuw, Gerard (Eds.), Problems of Excavating Cybernetics and Systems. BKS+, Southsea.
12. Glanville, R., 2000. The value of being unmanageable: variety and creativity in cyberspace. In: Eichmann, Hubert, Hochgerner, Josef, Nahrada, Franz (Eds.), Netzwerke. Falter Verlag, Vienna.
13. Glanville, R., 2009. The Black Boox Vol. III: 39 Steps, Echoraum; Vienna. Glanville, R., 2003. Machines of wonder. Cybern. Hum. Know. 10 (3,4), 91–105.

14. Hopf, S., Meier, N., 2011. Plattenbau Privat. 60 Interieurs, Nicolai, Berlin.
15. Pruckner, M., 2002. 90 Jahre Heinz von Foerster. Die Praktische Bedeutung seiner Wichtigsten Arbeiten (DVD), Malik Managment Zentrum, St. Gallen.
16. Sander, K., 1999. In: Sander, K. (Ed.), Heinz von Foerster: 2 2= grün (CD), supposeé, Küln. Scherbius, A., 1928. Ciphering Machine. US Patent No. 1.657.411.
17. Tessmann, O., 2008. Collaborative Design Procedures for Architects and Engineers (Ph.D. thesis), University of Kassel, Books on Demand, Noderstedt.
18. Thompson, D.W., 1992. On Growth and Form, The Complete Revised Edition Dover Publications, New York.
19. Turing, A.M., 1937. On computable numbers, with an application to the Entscheidungsproblem. In: Proceedings of the London Mathematical Society, vol. 42, Ser. 2, pp. 230–265.
20. Wilkinson, L., 2005. The Grammar of Graphics, 2nd ed. Springer, New York.
21. Von Foerster, H., 1950. Quantum mechanical theory of memory. In: von Foerster, H. (Ed.), Cybernetics: Transaction of the Sixth Conference, Josiah Macy Jr. Foundation, NY.
22. Von Foerster, H., 1970. Molecular ethology, an immodest proposal for semantic clarification. In: Ungar, G. (Ed.), Molecular Mechanisms in Memory and Learning. Plenum Press, New York, pp. 213–248.
23. Von Foerster, H., 1972. Perception of the future and the future of perception. Instr. Sci. 1 (1), 31–43.
24. Von Foerster, H., 1984. Principles of self-organization in a sociomanagerial context. In: Ulrich, H., Probst, G.J.B. (Eds.), SelfOrganization and Management of Social Systems. Springer, Berlin, pp. 2–24.
25. Von Foerster, H., 2003. Understanding Understanding. Springer, New York.
26. Weinberg, G.M., 2001. An Introduction to General Systems Theory. Dorset House,

CITATION

Thomas Fischer, Circular causality and indeterminism in machines for design, Frontiers of Architectural Research, Volume 3, Issue 4, December 2014, Pages 368-375, ISSN 2095-2635, http://dx.doi.org/10.1016/j.foar.2014.06.003.

CHAPTER 4

Micro- and Nano-Air Vehicles: State of the Art

Luca Petricca, Per Ohlckers, and Christopher Grinde

Department of Micro and Nano Systems Technology (IMST), Vestfold University College, P.O. Box 2243, 3103 Tønsberg, Norway

GLOSSARY

Infrared: The part of the invisible spectrum that is contiguous to the red end of the visible spectrum and that comprises electromagnetic radiation of wavelengths from 800 nm to 1 mm.

Aerodynamics: It is a sub-field of fluid dynamics and gas dynamics, and many aspects of aerodynamics theory are common to these fields.

ABSTRACT

Micro- and nano air vehicles are defined as "extremely small and ultra-lightweight air vehicle systems" with a maximum wingspan length of 15 cm and a weight less than 20 grams. Here, we provide a review of the current state of the art and identify the challenges of design and fabrication. Different configurations are evaluated, such as fixed wings, rotary wings, and flapping wings. The main advantages and drawbacks for each typology are identified and discussed. Special attention is given to rotary-wing vehicles (helicopter concept); including a review of their main structures, such as the airframe, energy storage, controls, and communications systems. In addition, a review of relevant sensors is also included. Examples of existing and future systems are also included. Micro- and nano-vehicles with rotary wings and rechargeable batteries are dominating. The flight times of current systems are typically around 1 hour or less due to the limited energy storage capabilities of the used rechargeable batteries. Fuel cells and ultra-capacitors are promising alternative energy supply technologies for the future. Technology improvements, mainly based on micro- and nanotechnologies, are expected to continue in an evolutionary way to improve the capabilities of future micro- and nano air vehicles, giving improved flight times and payload capabilities.

INTRODUCTION

Recently, a large number of studies on micro- and nano air vehicles (MAVs/NAVs) have been published [1–5]. MAVs are defined as small flying systems which are designed for performing useful operations [1]. In 1997, DARPA started a program called "MAV-project" where they presented some minimal requirements. In particular, they set the maximum dimension to be around 15 cm long, and the weight, including payload, to be less than 100 g [6]. Furthermore, flight duration should be 20 to 60 minutes. In addition to the MAV-project, DARPA started another program called nano air vehicles, which focus on the aim "to develop and demonstrate an extremely small (less than 15 cm), ultra-lightweight (less than 20 g) air vehicle system with the potential to perform indoor and outdoor military missions." [7].

In 2005, Pines and Bohorquez [8] published a review on the state of the art of unmanned air vehicles (UAVs), and many of the basic characteristics and challenges identified there are, to a large extent, valid also for NAVs, such as the challenges of maneuverability at low speed in confined spaces. In [9], NAVs are defined as small air vehicles with an operating range less than 1 km, a maximum flight altitude around 100 m, endurance less than one hour, and maximum takeoff weight (MTOW) of 25 g while MAVs are defined as 5 kg MTOW with endurance around 1 hour and an operative range around 10 km.

In this paper, we will use the definition from [9] when referring to MAV and NAV. When referring to both classes of systems, the term AVS (air vehicle systems) will be used.

Research and Development of AVS

Many research institutions are actively studying and developing new air vehicles, reducing size and weight while improving performance, and adding more functionality. Examples here are Harvard Micro-robotics Laboratory in the USA [10], Department of Aeromechanics and Flying Engineering from Moscow Institute of Physics and Technology in Russia [4, 11], Aircraft Aerodynamics and Design Group at Stanford University (USA) [12, 13], the Autonomous Systems Laboratory at ETH Zurich (Switzerland) [14, 15], and Deptment of Precision Instrument and Mechanology at Tsinghua University in China [16]. Several companies and agencies also play an important role in the manufacturing and development of AVS. Examples here are DARPA [7] from USA, Prox Dynamics [2, 17] from Norway, and Syma from USA.

Applications

AVS applications span a wide range, and the majority of them are military. AVS are capable to perform both indoor missions and outdoor missions in very challenging environments. The main applications are intelligence, surveillance, and reconnaissance (ISR) missions. These systems can provide a rapid overview in the area around the personnel, without exposing them to danger. Infrared (IR) cameras can give detailed images even in the darkness. Furthermore, NAVs, thanks to their reduced dimensions, are perfect for reconnaissance inside buildings, providing a very useful tactical advantage. As reported in [3], such small vehicles are currently the only way to remotely "look" inside buildings in the battlefield.

They can carry specific sensors such as gas, radiation or other sensors used to locate biological, nuclear, chemical, or other threats. They can, for instance, fly inside toxic clouds and transmit data or bring samples back to the base station, and, thus, provide vital information on the composition and extent of gaseous clouds and improve the assessment of danger.

Some of the applications described above can be extended to the civilian field. For example, the police and the fire brigade could use the capability of indoor flights for inspecting unsafe or collapsed buildings [2] in order to search for survivors or simply do a safety check of the building structure.

Since AVS would decrease the time necessary to explore a given area [3], they could be used in disaster cases, such as earthquakes, after hurricanes, or in collapsed mines [1]. In these cases, locating survivors faster increase the probability of saving lives.

However, AVS are not only related to high-risk applications, they can also be used as a support in regular police operations such as traffic control [1], crowd management or ordinary city surveillances.

Mass production of AVS will reduce the cost and, thus, enhance distribution among soldiers and policemen. This could render NAVs to be a natural part of the standard soldiers' equipment. In this case, one of the main features that NAVs must have is that they have to be ready for flight in a few seconds, without any lengthy startup procedures needed.

CHALLENGES

AVS are not only scaled down versions of larger aircrafts "they are affordable, fully functional, militarily capable, small flight vehicles in a class of their own" [6]. With their reduced size, they have to keep all the features of larger aircraft in a small volume, which increase the complexity and challenges. However, in the last few years, the miniaturization progress of AVS has practically stopped [4] (See Figure 1). This mainly happened since there are several problems. There are both physical and technological challenges that slow down further miniaturization [4].

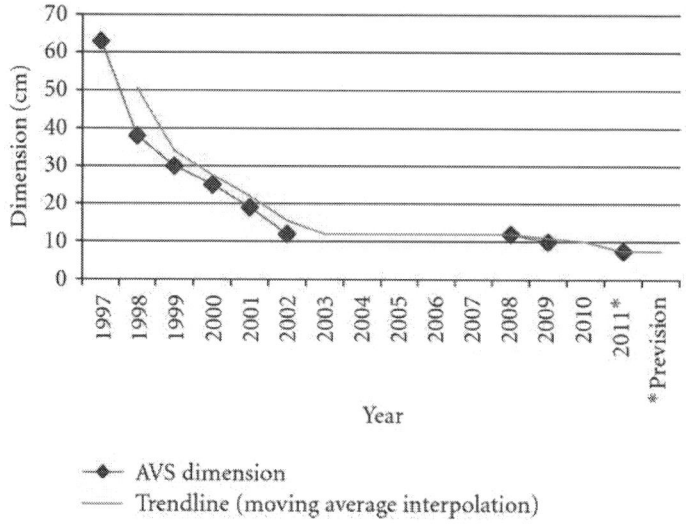

Figure 1. AVS development: dimension reduction; data from 1998–2002 [4] and 2008-2009 [2].

CHALLENGES

The first problem that appears is related aerodynamics related to the low Reynolds number for AVS. This dimensionless number reflects the ratio between the inertial forces and the viscous forces and is defined [18] as

$$\text{Reynolds number} = \frac{\text{fluid density} \times \text{speed} \times \text{size}}{\text{viscosity}}. \quad (1)$$

For AVS, both speed and size are several orders of magnitude smaller than for large aircrafts. This gives Reynolds number, less than one-hundred thousand, which is less than one-tenth of what is common for a full-size aircraft (Figure 2). Flight in this aerodynamic domain is more difficult. Since other physical laws are governing in this domain, a lot of efforts have been made to understand ultralow Reynolds number flight, studying the flight of insects whose size is even smaller than NAV.

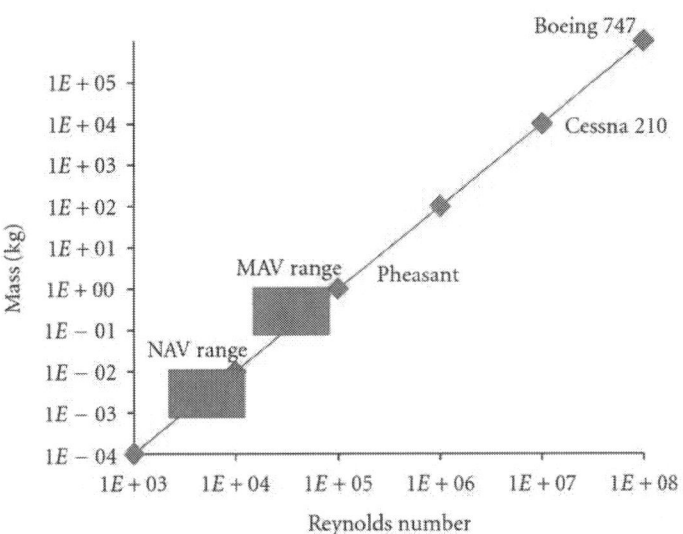

Figure 2. Reynolds number for aerial vehicles, adapted from [19].

Although aerodynamics at low Reynolds numbers are not clearly understood yet [5], it is well know that for Reynolds numbers under 100,000, the aerodynamics efficiency (defined as lift-to-drag L/D ratio) rapidly decreases [19, 20] (Figure 3).

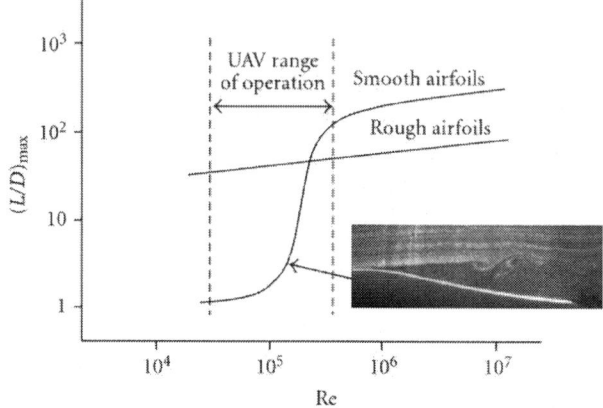

Figure 3. Lift-to-drag ratio variation with Reynolds number, reproduced from [73].

In addition to the physical challenges, given by the intrinsic reduction of physical parameters, there is also a problem of system integration. One can easily be misled to believe that larger aircrafts are much more complex than small AVS. The complexity of AVS becomes apparent if it is considered that they, similar to a larger aircraft, should be fully operational with respect to flight altitude, acceleration, stability, speed, and so forth, while the sensors and signal processing units, as illustrated in Figure 4, have to be integrated in a much smaller volume, with limited weight while keeping the power consumption to a minimum, increasing the challenges beyond that of larger aircrafts.

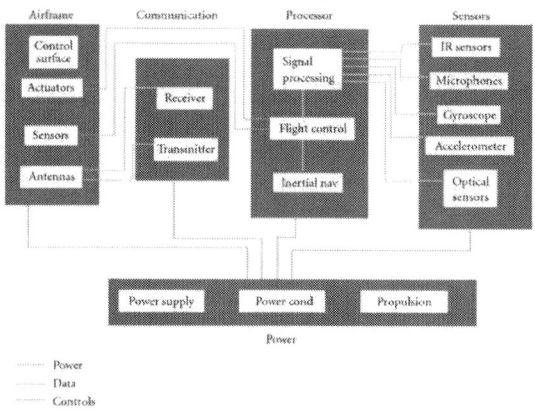

Figure 4. AVS system integration, adapted from [6].

Weight Budget and Power Budget

During the design process of AVS, both the weight budget and the power budget should be carefully monitored. In particular, the total mass of the vehicle should be kept as low as possible, since added weight will increase power consumption. The minimum power required to keep a fixed-wing aircraft in level flight can be expressed as [13]

$$P = \frac{TV}{\eta_p} = \frac{W}{L/D} \frac{(2W/S\rho C_L)^{1/2}}{\eta_p}, \qquad (2)$$

where T is the thrust, V is the velocity, η_p is the propeller efficiency, W is the weight, S is the wing area, ρ is density, and C_L is the lift coefficient. This means that doubling the weight nearly triples the power consumption. Similarly, for hovering flight, the power requirement is expressed as [13]

$$P = \frac{TV_h}{M} = \frac{W}{M}\left(\frac{W}{2S\rho}\right)^{1/2}, \qquad (3)$$

where M is the figure of merit of the rotor and V_h is the induced velocity in hover. Similar to that described above, a doubling of the weight increases the power required by a factor of nearly 3.

An example weight budget for a 197 g MAV can be found in [15]. The details are presented in Figure 5.

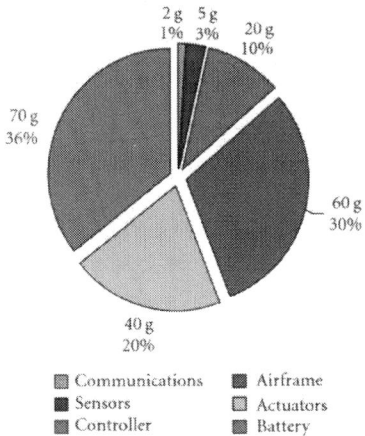

Figure 5. Weight budget for 197 g AVS as presented in [15].

A similar budget for a NAV can be found in [12], in which a 15 g vehicle is presented. However, a super capacitor rather than a battery was used as a power source since no other technologies were able to "satisfy the power requirement within the weight constraint". Since the power density of super capacitors at present is lower than that of batteries (see Section 5.3), the performance falls outside the specification for NAV. A more realistic comparison is, therefore, to replace the 5 g super capacitor with a 6 g battery as used in [2, 12]. The modified weight budget for a 15 g AVS is, then, as illustrated in Figure 6.

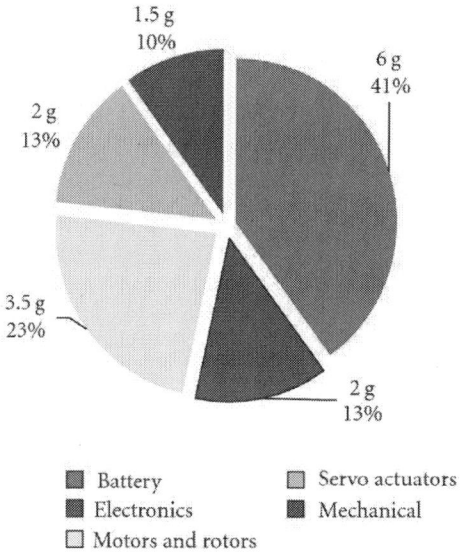

Figure 6. AVS weight budget allocation: the first number is the weight in grams while the second one is the percentage with respect to the total weight of 15 g, adapted from [2, 12].

It is interesting to see how the contributions from the various parts scale when the overall weight is reduced. A comparison of the information in Figures 5 and 6 after classifying the various parts in four categories (electronics, motors, battery, and airframe) can be found in Figure 7. It reveals that if the size is decreased, electronics still account for about 13% of the total weight, while motors, actuators and battery increase relatively. This reflects the difficulties of scaling down batteries and motors while maintaining acceptable performance.

CHALLENGES

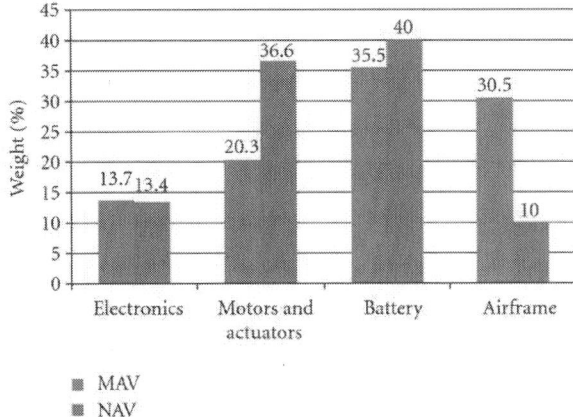

Figure 7. AVS weight budget allocation in percentage with respect to the total weigh calculated for a 197 g MAV and 15 g NAV.

When the system is miniaturized, the airframe has the greatest reduction in percentage. This is probably due to the ultra limited weight budget of the NAV that made the developer really optimize the airframe, carefully selecting shape and materials.

A similar comparison has been made for the power budget (Figure 8). It should be noted that, in this case, the MAV power budget has been adapted from [15] while for the NAV budget, we did not find any paper which explicitly reported such information, and for this reason, it was estimated. We used the theoretical estimate for a 15 g NAV found in [21]. For an additional margin to account for various system losses, the value was increased by ~15%, for instance, assuming an efficiency of the electric motor of 85% [22]. The required power was, therefore, increased from the reported 585 mW to 700 mW. With respect to the communication systems, the authors in [16] report an example of a handmade RF communication apparatus that weighs 8 g, which we assume could be adapted to be used in an NAV. The power consumption of this transmitter is reported to be around 500 mW. For the remaining onboard devices, among them a camera [23], the power requirement was estimated to be about 50 mW. Comparing the corresponding power budgets, as seen in Figure 8, we find that by downscaling the weight, the NAV at less than one-tenth of the weight of the MAV, relatively uses less power to generate lift. Since the power consumption of the communication system is dependent on factors like distance, bit rate, data compressions, and so forth, rather than size, the relative requirement in the NAV is significantly higher.

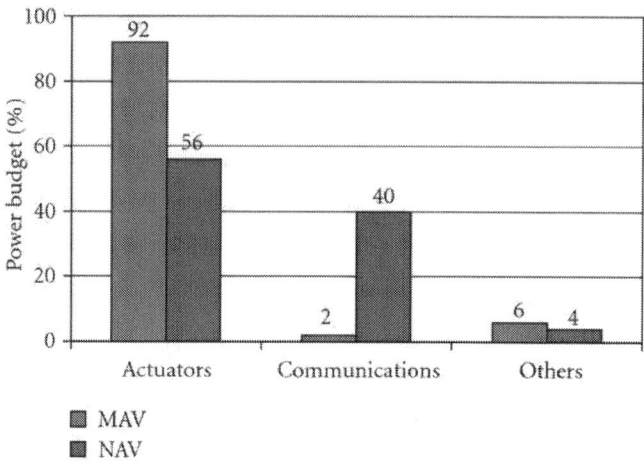

Figure 8. AVS power budget allocation in percentage calculated for a 197 g MAV and 15 g NAV.

AVS TYPOLOGIES

AVS can be classified into four main typologies depending on their method of propulsion and lift. These are fixed wings, rotary wings, and flapping wings. The fourth class is without propulsion and is called passive.

In the followings sections, we will briefly review one or more examples from each class, and analyze their main advantages and disadvantages.

Fixed Wings
Among the different typologies of AVS, fixed wing is the most developed and the easiest to design and build, because "well-established design methods for larger operational fixed-wing UAV could be applied with some precautions and modified aerodynamic characteristics" [20]. Several prototypes have already been proposed to customers [1]. These kinds of vehicles require relative high speed for flight, typically 6 to 20 m/s. As they are incapable to hover or fly any slower, indoor flight is very challenging and is often avoided. Examples of suitable applications are location of forest fires, searching for people at sea, and missions where low speed is not required.

Fixed-wing aircrafts require a thrust-to-weight ratio less than 1, since wings provide an additional lift [13]. The full weight is divided by the L/D ratio, and, thus, they require less power to fly than a helicopter with the same weight hovering, where the weight is completely balanced by the propulsion thrust [13]. This efficiency gain is most obvious in larger aircrafts where the L/D ratio reaches values of more than 30. Unfortunately, as we explained in the Section 2, this parameter rapidly decreases as the dimensions and correspondingly, the Reynolds number decrease (Figure 3). For this reason, the obvious advantage of a large aircraft becomes less pronounced when the ratio L/D is reduced to less than 10. Several prototypes exist, but none are in the NAV range [9]; the existing AVS models have wingspans larger than 15 cm and thus are considered MAVs. In [16], two examples of fixed-wing MAVs are shown. One of these, the TH360 (Figure 9) includes a color video camera that transmits real-time images to the control station. The propulsion is from an electric motor, the wingspan is 45 cm and the total weight is around 120 g.

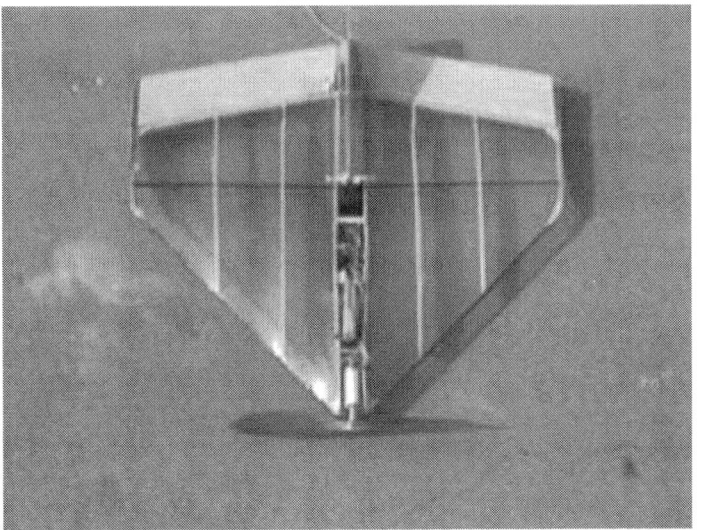

Figure 9. TH360 prototype MAV [16].

The main drawback of the system is the limited 5-minute flight endurance. Another fixed-wing MAV, often referred to, is the black widow [24] which was developed in 2001 by AeroVironment Inc. in collaboration with DARPA. The performance of this MAV is reported in Table 1.

Table 1. Performance summary for the first-generation Black Widow MAV [24]

Total mass	56.5 g
Loiter drag	9.4 g
Lift/drag ratio	6.0
Loiter velocity	11.2 m/s
Loiter lift coefficient	0.42
Loiter throttle setting	70%
Endurance	33.4 min

Notice that the flight endurance is much longer than for the TH360 in [16] discussed above. The L/D ratio is only 6 for this 15 cm wingspan MAV, which is less than one-fifth of typical commercial airliners.

Rigid wings were used in all of the examples presented above. In contrast, in [25], a flexible-wing MAV is reported and compared with a rigid thin-wing MAV of the same size and shape. The motivation for this comparison is "for highlighting the distinct aerodynamic advantages of the flexible wing." The conclusion is that a "deformable wing is expected to harvest an intrinsic benefit: a portion of the energy that would normally be lost to the wing-tip vortices and wake, downstream of the MAV, now is stored as elastic strain energy in the wing's structure" [25]. Following from this, it is found that flexible wings provide better L/D ratio than rigid wings for angles of attack (α') smaller than 10.

Rotary Wings

The second type of MAV typology are systems with rotary wings. These AVS basically have the same structure as macroscale helicopters and, thus, are able to fly at quite high speeds, hover, and execute vertical take-off and landing (VTOL). These features make them perfect for indoor flight and short-range reconnaissance. Due to larger power requirement for hovering and VTOL, the endurance is also the bottleneck for this kind of AVS. With the miniaturization, a lot of challenges arise. Examples are low efficiency of the rotor system and the low thrust-to-weight ratio [26]. Despite these disadvantages, rotary AVS are the only configuration capable to "combine acceptable high and low speed characteristic including hovering" [1]. Furthermore, they are also "the only controllably hovering object at the moment" [1].

Based on the number and position of the propellers, there are several possible configurations for rotary AVS. In Figure 10, some of the possible configurations are reproduced from [27].

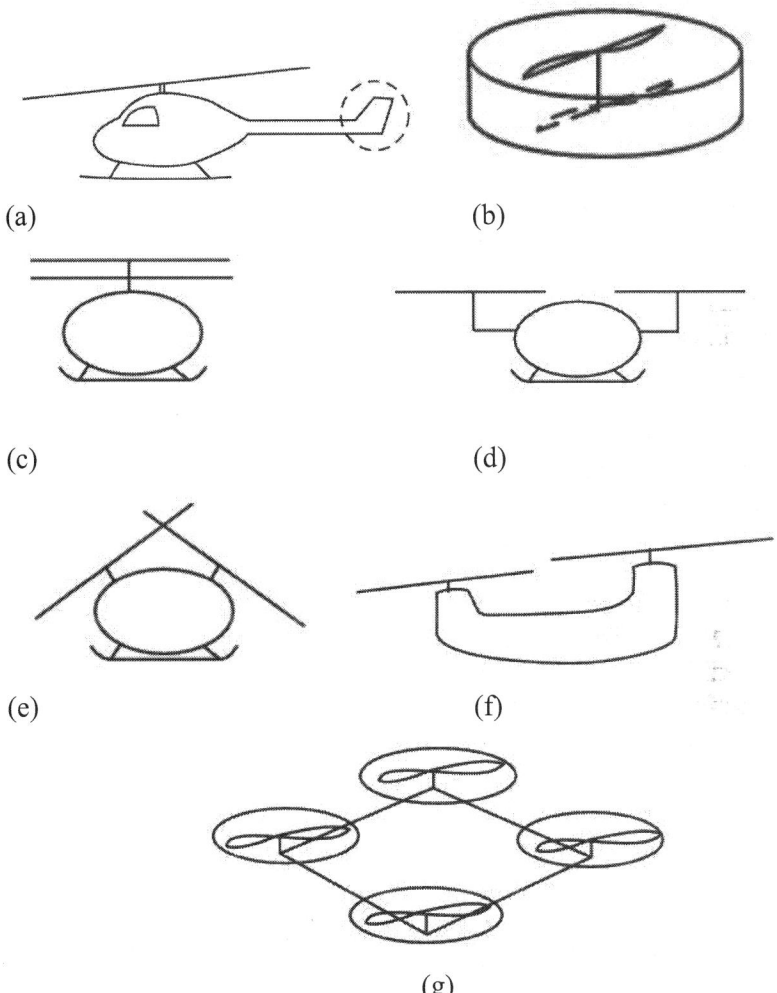

Figure 10. Graphic representation of rotary-wing configurations: (a) conventional configuration, (b) ducted coaxial, (c) conventional coaxial, (d) rotors side by side, (e) synchropter, (f) conventional tandem, (g) quad rotor [27].

Each one of these configurations has some specific features that make them suitable for specific types of missions.

Table 2 reports various aspects of the six typologies. Choice of configuration will depend upon the mission requirements. For example, if one wants to build AVS that are easy to maneuver, one should focus more on quad-rotor configuration and discard coaxial, but if we need AVS with low structural complexity, then quad-rotors are not the optimal choice anymore.

Table 2. Selection criteria of different rotary AVS typologies [27] (1: bad, 10: very good)

Selection criteria	Conventional (a)	Ducted coaxial (b)	Coaxial (c)
Compactness of folding	1	2	10
Reliability	9	10	8
Controllability	5	5	7
Aerodynamic cleanliness	8	2	8
Maturity of technology	10	8	10
Hover efficiency	10	8	8
Aerodynamic interaction	7	7	7
Vibration	1	2	1
Cruise efficiency	7	6	8
Maneuverability	5	3	3
Ease of payload packaging	10	8	10
Simplicity of structure	8	8	10
Simplicity of control system	6	6	6

Selection criteria	Side by side (d)	Tandem (f)	Quad-rotor (g)
Compactness of folding	2	2	8
Reliability	8	8	9
Controllability	5	5	10
Aerodynamic cleanliness	6	6	2
Maturity of technology	9	10	5
Hover efficiency	10	10	8
Aerodynamic interaction	10	10	7
Vibration	2	2	2
Cruise efficiency	6	6	5
Maneuverability	4	4	9
Ease of payload packaging	10	10	8
Simplicity of structure	8	8	7
Simplicity of control system	6	6	10

In general, when a designer needs to choose the best configuration, he will use a combination between these different selection criteria. In [27, 28], a weight parameter is assigned to each selection criteria to identify which feature is more important.

Several prototypes of these configurations are reported in the literature. In 2004, Epson built a prototype of small ducted coaxial AVS [29].
This "flying robot" had a wingspan of 136 mm and a total weight, including batteries, of 12.3 g. At 3 min, the endurance was the weak point of this prototype.

AVS with an improved flight time of 10 minutes are described in [30]. However, compared to the AVS in [29], the AVS in [30] had a total weight around 110 g and only one rotor, and, in this case, the torque is balanced by changing the angles of the yaw control surfaces [30], so these two systems are not directly comparable.

The T-REX model, provided by ALIGN Corporation [31], is a good example of a conventional configuration rotary MAV. It is an electrical powered AVS with rotor diameter less than 50 cm and a total weight around 340 g.

Despite that this configuration is the most common for larger aircrafts, it suffers from the disadvantage that it is difficult to control with respect to the quad-rotors configuration. For this reason, a lot of research has recently been put into the control of AVS with quad-rotors. A basic study on control theory presented in [32] shows how quad-rotor AVS could easily be controlled by changing the rotation speeds of the motors. The same principle is used in [12, 14] and is illustrated in Figure 11. The torque is cancelled by making two rotors rotating clockwise and two rotors rotating counterclockwise. Despite the simple control systems and the ease of maneuverability of the quad-rotor systems, there are two main disadvantages that could limit their success. In particular, quad rotors mean four motors that are very power consuming [14]. Furthermore, motors are, in general, heavy and difficult to miniaturize. For these reasons this configuration should be avoided when the main target is to maximize flying time.

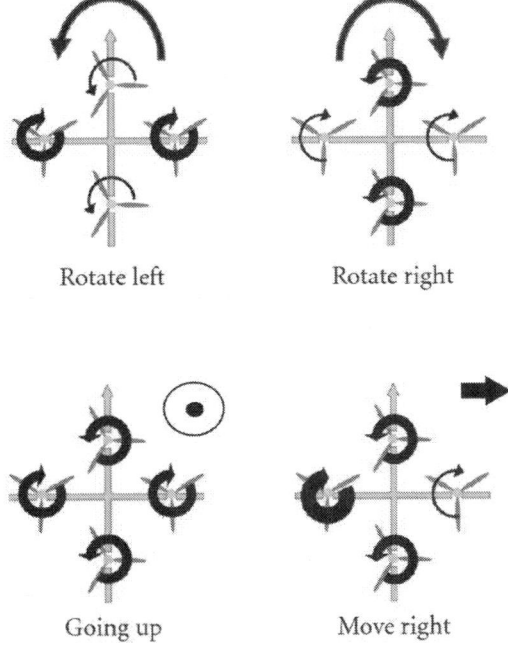

Figure 11. Quad-rotor concept motion description, the arrow width is proportional to propeller rotational speed adapted from [14].

Flapping Wings

Both fixed wings and rotary wings provide mature and well-known technologies, but have problems due to the high unsteady effects due to reduction of Reynolds numbers. This motivated researchers to investigate alternative typologies. The basic idea was adapted from nature and uses the same flying technique as insects and birds: flapping wings. Since this idea came up, a lot of studies have been done in order to investigate the efficiency of such methods and the possibility to reproduce them in the laboratory. In fact, the principal motivation seems to be the possibility to "integrate lift and thrust together with stability and control mechanism" [33]. However, when we refer to this class of vehicles, we should make a distinction between bird-like vehicles called ornithopters and insect-like vehicles called entomopters. These two subclasses of flapping wings have completely different features. Ornithopters, like the majority of birds, generate lift by flapping wings up and down with synchronized small variations of angle of incidence. This method of thrust generation require forward flight similar to fixed-wing AVS [1]. As a result, ornithopters cannot hover, and they need to obtain an initial airspeed before taking off [1]. Entomopters use the kinematics of insects for flying, meaning a "large and rapid change of angle of incidence" [1]. Due to this large angle variation between the upstroke and down stroke, this technique is sometimes also referred to as pitch reversal. Compared to how birds fly, they are able to generate much more lift and, thus, are able to execute VTOL [1] and hovering.

With these two advantages, entomopters are much more interesting to adapt to AVS than ornithopters, and therefore we will mainly discuss Entomopters. Insects generate wing beats by contraction of muscles. The muscles can either generate wing motion through direct attachment or through indirect attachment, where the wing motion is generated, for example, through deforming the shape of the thorax [34, 35] (Figure 12). Large insects with lower beat frequency use both these modes of flying while smaller insect use mainly indirect muscles [35].

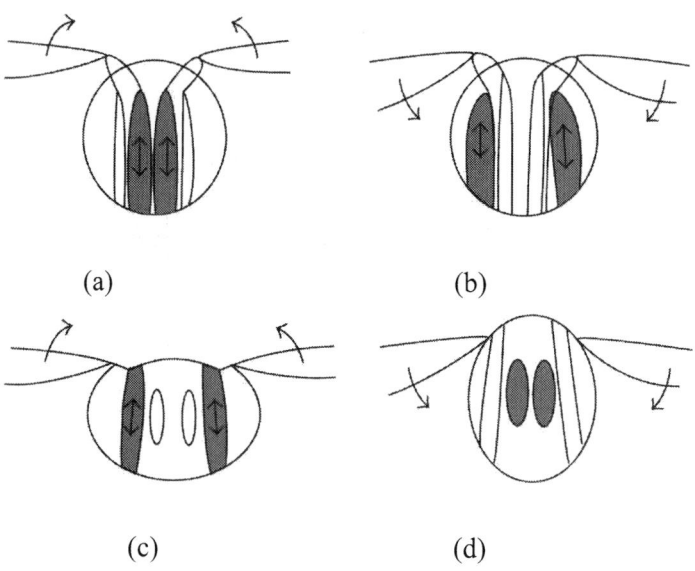

Figure 12. Insect wings beat generation: direct muscles (a, b) and indirect muscles (c, d) [35].

Linear actuators are the most suitable means for reproducing this motion. However, none of the available actuators match natural "muscles over all the main performance characteristics" [35]. The most promising technology is electro active polymers (EAP) that is able of providing high energy density and high efficiency. In [36, 37], the possibility of using EAPs to reproduce artificial muscles is investigated. Even though it looks very promising, this technology is still under development and thus not yet widely available in commercial actuators.

A possible concept is to develop a flapping-wings mechanism using rotary actuators such as electric motors. In [34, 35, 38], three different methods for translating rotary motion into flapping-wings motion are presented (one example is shown in Figure 13). All three uses a crank rocker technique, but differ with respect to how they generate the pitch motion of the wing.

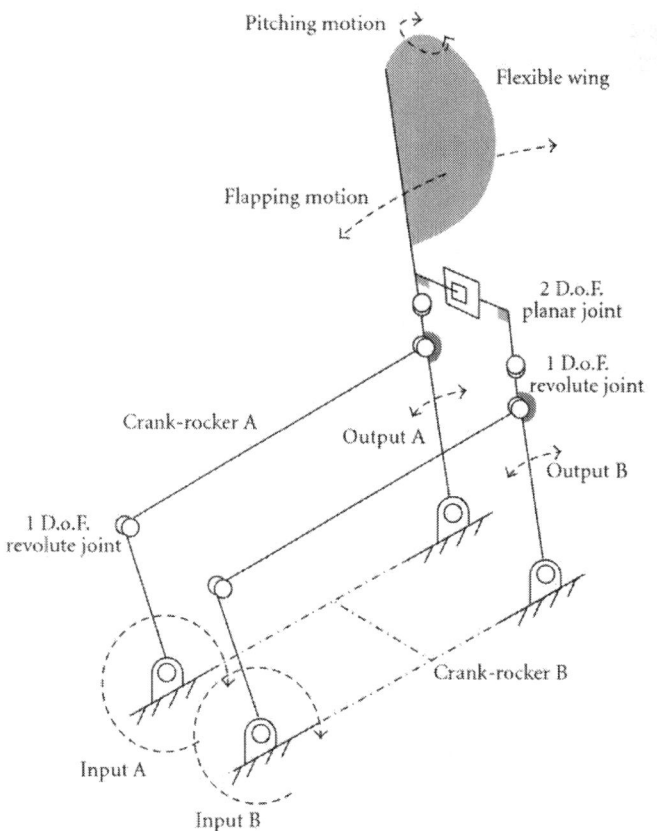

Figure 13. Rigid-body representation of the parallel crank-rocker mechanism. Full revolutions at inputs A and B produce reciprocating outputs A and B, which cause the wing to flap. Introducing a phase lag between the two linkages (and hence outputs) introduces a pitching motion to the wing [35].

Despite this, a recent study [39] has shown that it is also possible to generate stable forward flight using simple motion without any feedback control of the wing movement.

Another important aspect of flapping air vehicles is the shape of the wings. In the literature, many different designs are reported; however, most of them try to reproduce bird or insects wings.

One distinction can be made between smooth wings and rough wings. Several test runs by the authors of [39] show that creating rough wings helps to increase the performance, since they "are needed to prevent feathering deformation and produce a large up-down body motion" [39].
Regarding wings, they are in general fabricated using micro machined molds. The molded wings are in general made from polymers like thermoset resins [39, 40].

Other Techniques

While the three typologies described above are the most common, some authors have started to investigate alternative techniques that could be useful in some cases. The most important class are the passive AVS. In [10, 41], a "palm-sized autonomous glider" (Figure 14) with a 10 cm wingspan weighing only 2 g is described. Since this class of vehicles does not have generation of thrust, meaning they are passive, they need to be hand launched or dropped, for example, from aircrafts.

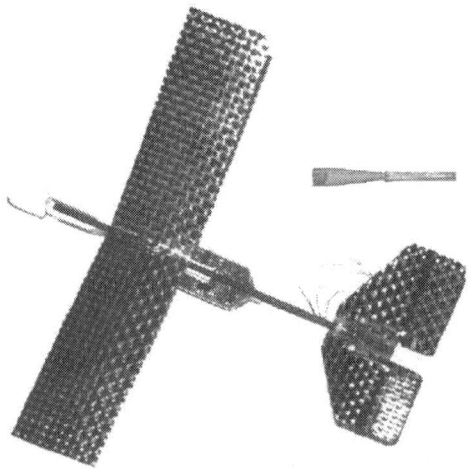

Figure 14. Palm-size micro glider [41].

Another example of an alternative topology is a blimp or airship. Although an unmanned airship is presented in [42], the dimensions are out range of the specification of NAV and MAV.

AVS TYPOLOGIES COMPARISON

Among the different typologies presented, it is important to choose one that offers the greatest advantages in terms of reliability, efficiency, and suitability for the application areas presented in Section 2. Unfortunately, none of these typologies have advantages only, and thus compromises must be made. Since fixed-wing vehicles are able to fly with much less thrust with respect to their real weight, they are perfect for relative large area control and high autonomy missions. Unfortunately, as we reported in Figure 3, the L/D ratio sharply decreases as soon as the Reynolds number decreases below 500000, and, thus, the main advantages of this typology (low consumption) become less pronounced compared to larger aircrafts. Since the Reynolds number decreases rapidly with miniaturization, the obvious advantages of the fixed-wing topology compared to other typologies such as helicopters, becomes less pronounced.

Also, since fixed wing topologies are incapable of hovering, they are less suitable for finding and overseeing targets. In [43], an algorithm that allows fixed-wing vehicles to lock on to the target while flying is investigated. The results are still unsatisfactory and too complex. In addition, their lack of hovering capabilities also renders them less suitable for indoor missions. For many customers, such as military and tactical squads, this limitation is not acceptable.

For users that require hovering capability, rotary-wing typology appears to be a good solution. In addition to hovering, they offer good maneuverability and medium complexity.
This is also the only configuration capable to "combine high and low speed characteristics [1].

The main disadvantage of rotary wings is the relative high power consumption. To overcome this problem, many researchers have investigated flapping wings. The basic idea is that if insects have this kind of propulsion, and they are able to generate high lift, they are probably very energy efficient. Unfortunately, this is challenging, since nature and engineering are based on different evolution processes. Nowadays, it is still not fully documented whether this method provides better efficiency than the much simpler rotary wing principle [13]. Some authors [38, 44] claim that they provide better performances, others [11] claim the contrary. However, these studies are not very comprehensive and should be considered with caution. In [21], a comparison between a flapping-wing system and a rotary-wing system is done. It is reported that the hover efficiencies only differ with less than 5% and thus are almost negligible. Therefore, it can be assumed that both principles have comparable efficiency in the ultralow Reynolds number regime. However, rotary-wing systems are simpler and can, therefore, more easily be miniaturized. In addition, "they are also the only controllably hovering flying objects at the moment" [1].

For this reason, rotary wings are acknowledged as the most suitable topology for NAV [11, 13, 21]. Table 3shows one comparison table reproduced from [14].

Table 3. Flying principle comparison focused on ability to miniaturization [14] (1: bad, 2: medium, 3: good)

	Fixed	Rotary	Bird	Blimp
Power cost	2	1	1	3
Control cost	2	1	1	3
Payload	3	2	2	1
Maneuverability	2	3	3	1
DOF	1	3	3	1
Stationary fly	1	3	2	3
Low-speed fly	1	3	2	3
Vulnerability	2	2	3	2
VTOL	1	3	2	3
Endurance	2	1	2	3
Miniaturization	2	3	3	2
Indoor usage	1	3	2	2
Total	20	28	26	26

ROTARY WINGS: MAIN PART ANALYSIS

In this section, conventional helicopter NAVs are reported and analyzed. Many of the discussions made in this section can be easily extended to the other typologies, since technologies such as energy storage, propulsion, and communications, are independent of the typology.

Airframe

It is well known in aerodynamics that the shape that provides the best aerodynamic performance in subsonic speed is the drop-based shape [45].
The design begins typically with finding placements for large components such as motors and battery. During this step, the center of gravity of the system and its location should also be considered. This space optimization is not an easy problem. In [46], an example of design optimization and positioning of the center of mass for a nano-helicopter is presented. In particular, a generic-based algorithm is used for "organizing a given set of components and payloads such that the resulting flight vehicle has the most compact overall size and still fulfills the given physical and control constraints" [46]. Once all the parts of the system have been allocated, it is possible to start to design the enclosures of the vehicle as a double drop-shaped hull (Figure 15).

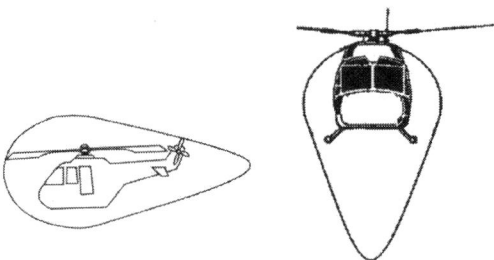

Figure 15. Double drop section: lateral view (left) and vertical section (right).

The horizontal drop shape (Figure 15(left)) is needed to reduce the aerodynamic drag during forward flight. Since the main rotor pushes down the air, it will create an airflow that pushes the helicopter body towards the ground, increasing the power consumption. Once more, a drop-shape body cross-section (Figure 15(right)) will reduce the drag forces, reducing the power consumption further. Airframe design can truly be considered much more than an aesthetic endeavor. Due to the lack of useful simulators in the ultralow Reynolds number regime, many companies base their development more on workmanship experience than on simulations.

Carbon fiber composites are the main materials used for AVS airframes, because they have high strength-versus-weight ratio and are easily accessible.

Propulsion

All the three active AVS typologies presented above need to generate motion. There are several ways to make an air vehicle fly. The most common and easiest way is to use electric motors. Almost all the existing prototypes use electric motors due to their high efficiencies, reliability, and ease of control.

Since coreless motors are lighter and smaller than, for example, direct current (DC) iron-core motors, they are considered more suitable. As their name indicates, there is no iron core inside their motor structure. The magnet is positioned directly inside the coil, and then the rotor coil is wrapped around the magnets without using any iron material as illustrated in Figure 16.

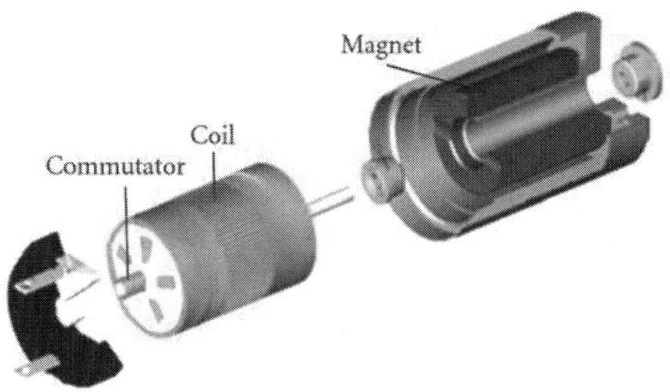

Figure 16. Coreless motor structure, from [74].

In addition to the small dimensions and the low weight, another advantage of coreless motors is the lack of iron losses that are reflected in a higher efficiency. Furthermore, since the rotor is very light, it has a small inertia that allows extremely fast accelerations and decelerations.

However, the lack of iron in the center reduces the motor heat dissipation. To avoid overheating and thermal problems, they are only used for small and low-power motors.

Although an electric motor is the most suitable for AVS applications, it can be seen in Figure 8 that more than the half of the electric energy present in the AVS is used to generate lift. It could, therefore, be advantageous to replace electric motors with other systems, such as gas turbine or internal combustion engines (ICE). Two AVS examples using ICE are presented in [47], in which a motor provided by Cox Company were used. However, as the engine of this vehicle was quite large, it will be difficult to integrate into future NAVs. Most recent studies on miniaturized ICE [48] show that, with the current technology, it is possible to build a miniaturized combustion motor of 0.3-0.4 cc.

Even though ICE motors are interesting for AVS applications; they suffer of one big disadvantage compared to electric motors: they are very noisy. This limits the application in NAVs that have to be used for tactical missions, where it is required to have high stealth skills.

Micro-gas turbines could be another alternative. In [49], an example of the fabrication of an extremely small turbine with dimensions around (2cm×2cm×0.4cm) with a combustion chamber of 0.195 cubic cm is given. It has been built using six silicon wafers and configured for hydrogen fuel (Figure 17).

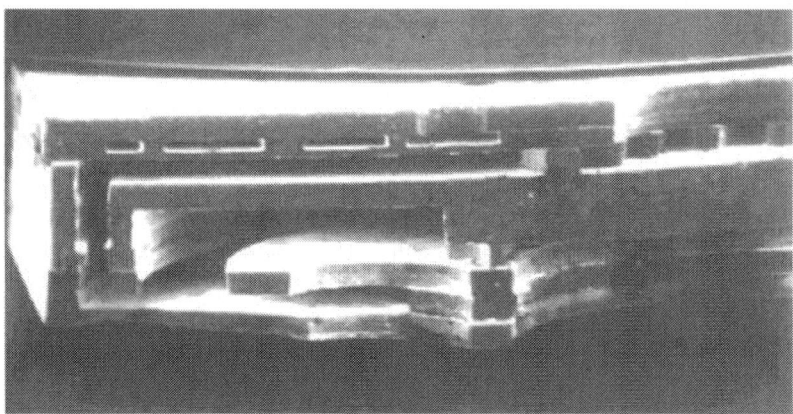

Figure 17. Micro gas turbine, SEM image [49].

ONERA Company announced in 2008 that they had demonstrated a micro gas turbine suitable for AVS. Their turbine can supply from 50 to 100 W with dimensions around 2-3 cm for the diameter and the height [50]. The combustible in this case could be either hydrogen or propane (Figure 18).

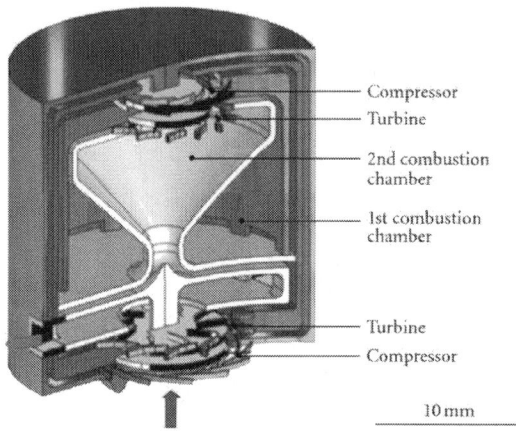

Figure 18. Architecture of the ONERA Micro gas turbine, reproduced from [50] ONERA, the French Aerospace Lab.

Despite significant efforts by numerous groups [49, 51], no known commercialization of MEMS gas turbine generators are currently known.
Finally, the possibility to include a hybrid system, such as electric motors and combustion engines should be mentioned. Although this technique has been already used with good results in larger air vehicles [52], it is not suitable for smaller systems such as NAV in which both weight and size represent very strict constraints.

Energy Storage

All the active AVS typologies need electric energy on board to feed the electronic circuits, sensors, actuators, and the communication devices. Furthermore, if electric motors are used, a large part of the electric energy will be used for the propulsion supply motors. Since energy stored in batteries does not require any conversion to be useful for both the electronics and propulsion, batteries seem to be the most appropriate for electric MAV. Furthermore, the energy density of the batteries has steadily increased during the last years, mainly thanks to the effort of the companies producing smart phones, notebooks, and other consumer electronics. In 2006, the battery market reached 50 billion dollars and is expected to expand to more than 70 billion in 2011 [53]. As a result, many companies continue to invest large amounts of money on battery research, mainly focusing on reducing dimensions and increasing the energy density.

Ni-Cd batteries have now almost completely been replaced by more energy dense lithium-based batteries that also are less toxic [53]. Furthermore, the most advanced batteries (intelligent batteries) include circuitry that optimizes the cells' discharge curves with respect to the loads. They "exploit various battery-related characteristics such as charge recovery effect, to enhance battery lifetime and ensure safe operation" [54].

Unfortunately, despite these improvements, the most advanced batteries also provide much lower energy densities than sources, such as gasoline or methanol, as shown in Figure 19.

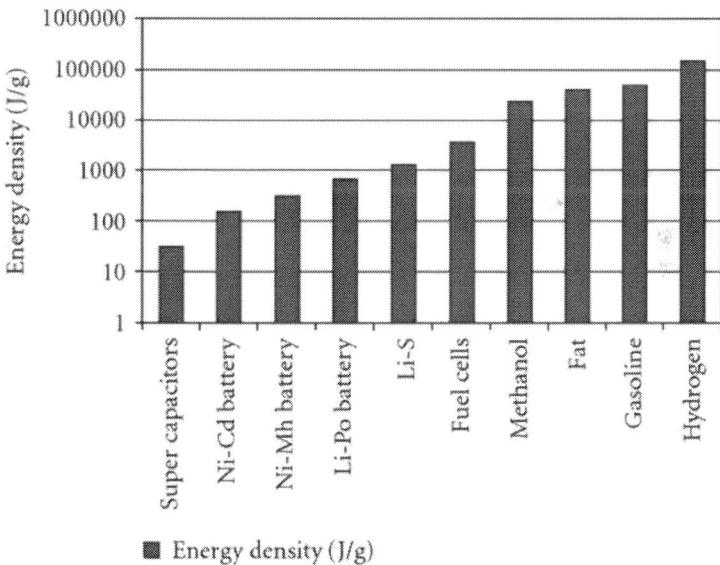

Figure 19. Energy densities of various energy storage systems, graph adapted from [33, 53, 75–77].

Methanol could be used in a gas turbine as presented in the previous section, providing much more power compared to battery based systems. Another alternative is fuel cells. A fuel cell system is conceptually a sort of battery in which the fuel is transformed into electric current trough an electrochemical process. There are several kinds of fuel cells which mainly differ with respect to the principle of energy conversion. Currently, the most promising fuel cells for AVS are proton exchange membrane (PEM) fuel cell and direct methanol fuel cell (DMFC) which could be considered as a subcategory of the PEM.

The schematic of a PEM fuel cell is reported in Figure 20 [51]. It basically consists of an anode catalyst and a cathode catalyst separated by a membrane that can be crossed only by protons. In the classical configuration, PEM uses hydrogen on the anode side and oxygen on the cathode side.

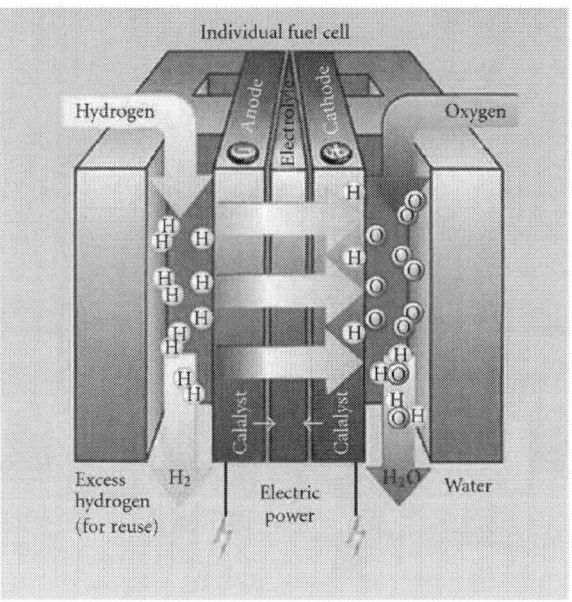

Figure 20. PEM fuel cell principle [78].

The hydrogen atoms that reach the anode catalyst dissociate into protons and electrons. The protons can cross through the membrane and reach the cathode. Here they will react with oxygen to form water. The electrons left behind the membrane are forced to "travel" inside the electric circuit, generating an electric current. It is clear that, in this case, the only residual waste is water, and thus this is very environmental friendly. Unfortunately, hydrogen does not occur naturally and thus has to be produced by chemical processes, wasting energy. Furthermore, hydrogen has a high mass energy density (143000 J/g) but a very low volumetric energy density (10790 J/L), which makes it difficult to store. In fact, it has to be transformed to liquid form, compressed at high pressure, or transformed to other forms with high volumetric densities, such as metal hydride.

To overcome this problem, several new kinds of cells have been developed. Among these, one of the most interesting is the DMFC. It uses pure methanol as fuel, which simplifies the fuel storage.

The working principle is similar to PEM, but the chemical reactions are a little bit more complicated. DMFC residual wastes are water and carbon dioxide, and the overall cell efficiency is lower compared to a PEM cell. These fuel cells are already available in the market [55], with quite small dimensions (several cm) and low weight (few hundred grams). An example of an MAV powered by a fuel cell is "The Hornet", developed in 2003 by DARPA, it "uses absolutely no batteries, capacitors, or other sources of energy" (except fuel cells) [56]. The Hornet has a wingspan of 38 cm and a total weight of the vehicle around 170 g, including fuel.

Since the weight budget of NAVs is very limited, further development is required before fuel cells become a viable alternative energy source.

Besides fuel cells, ultra capacitors have become interesting the last few years. The latest improvements have made this power storage principle attractive also for AVS applications, and they have already been used in some prototypes [12]. Since they are an evolution of normal capacitors, their main features are fast charging, high peak current, and virtually unlimited charge-discharge cycles [57]. The main drawback consists of the output voltage that strongly depends on the charge status of the capacitor. By the time an ultra-capacitor reaches a 25-percent state of charge, its voltage has dropped by half [57]. Besides this, when compared with other energy sources, they have a relative low energy density. Recently, in [58], a new electrostatic Nano capacitor (shown in Figure 21) which "dramatically increases energy storage density of such devices—by a factor of 10 over that of commercially available devices—without sacrificing the high power they traditionally characteristically offer" was announced.

Ultra capacitors are, therefore, candidates to play an important role for future energy storage systems in AVS. Solar cells should also be mentioned as a potential useful energy source. Even though photovoltaic systems are interesting for AVS, the small dimensions of NAVs, the weight budget constraints, and the mainly indoor application area (low light) limit the efficiency and the available energy.

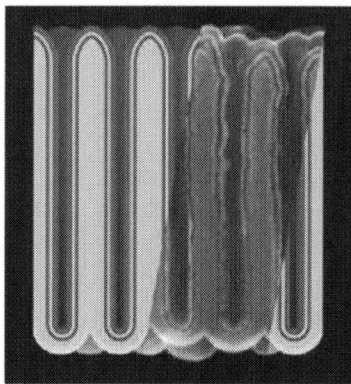

Figure 21. Electrostatic Nano capacitor. Developed by Maryland Nano Center [58].

Transmissions

Basically, for AVS, two different kinds of signals have to be transmitted: control signals and data signals. The control signals are needed for take-off, landing and for piloting the vehicle in general while the data signals are basically the data collected by onboard sensors of AVS, such as camera, microphones, gas sensors, and so forth, which for many applications need to be transmitted to a base station.

Control signals are mainly transmitted from the ground station to the vehicle while the data is sent from the vehicle to the user. An example of control communication systems on board is given in [16], where they developed a home-made RF transmitter for use onboard an MAV. With a weight of 8 g, it could transmit at 56 mW. The transmitter operated at a frequency range between 1.18 and 1.45 GHz. Furthermore, a micro demodulator operating at 50 MHz and weighing 5.4 g was used at the receiver end [16].

Another example is found in [25], where only a receiver was used onboard. In particular, the receiver, including a phase locked loop (PLL), weighed around 12 g, and it consisted of a 7-channel pulse code modulation system. Since no data was to be sent back to the base station no transmitter was required on board that vehicle.

When reducing the AVS dimensions, the major challenges for the communications parts are represented by the weight and size of the antennas, filters, and resonators. Antenna shape strongly depends on the operating frequencies and, thus, will depend on external factors, such as application (military frequencies are different from civilian frequencies), distances, bit rate, and so forth. This requires the antenna design to be application specific.

The size, weight, and performance of resonators depend on the operating frequency. Several examples of micromechanical resonators for various frequency ranges can be found in the literature. In Figure 22 is shown one example of it reproduced from [59]. Since quartz resonators suffer from high power consumption and relative bulky size [59], developing MEMS resonators could help overcome these drawback, and, thus, this is the most interesting candidate for substituting the quartz resonators [60].

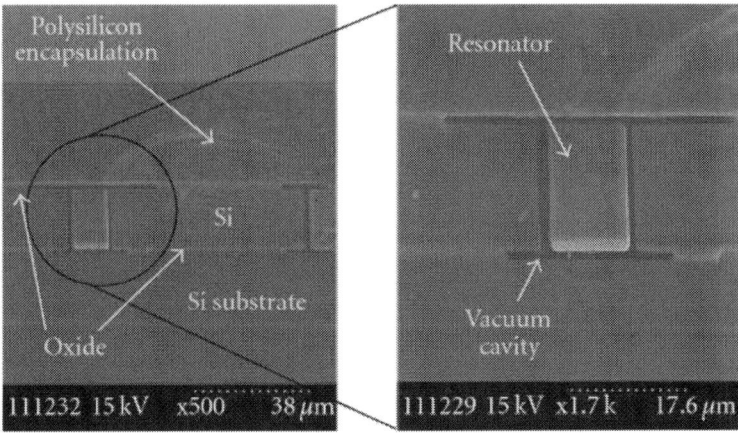

Figure 22. Example of an MEMS resonator [59].

Other examples of filters and resonators can be found in [61, 62], where systems for 22 GHz and 140 MHz bands were described.

Such devices can help to not only decrease the overall power consumption of the communications systems, but their small size and weight relative to the quartz systems they replace also help reduce the size, weight and power consumption of the overall system.

Sensor and Actuators
Sensors can roughly be divided into two categories. The first one contains the sensors that are necessary for flight control, the second is sensors that are a part of the payload and provide mission-specific information.

Theoretically, AVS should be able to fly only with a 3-D accelerometer and a 3-D gyroscope. Ideally, if we know the initial position, we will be able to calculate all the later positions only by integrating the resulting vector acceleration two times to find the position, while 3-D gyroscope signal is used to maintain flight stability. However, since all gyroscopes and accelerometers suffer from offsets and drifts, for instance with time and temperature, the accuracy of the calculated position will decrease over time. Additional sensors can be used to compensate somewhat for drifts and offsets. For example, in [63], it is stated that "accelerometers and gyros can only be used for the pitch and the roll, while for yaw measurements, magnetometers have to be used" [63]. In fact, if the roll rate is integrated with respect to time to find the roll angle, it "will lead to drifting errors" [63]. Another interesting solution is presented by the "Paparazzi project" described in [64], in which "a free and open-source hardware and software project intended to create an exceptionally powerful and versatile autopilot system" is described. The proposed solution uses two infrared (IR) sensors positioned on the side walls of the AVS.

The basic concept is that if the vehicle is perfect parallel to the earth surface, then both the sensors will detect the same temperature and thus get the same signal. On the other hand, if there is any misalignment (Figure 23), one sensor will reveal the earth temperature (warmer) while the other one will reveal the sky temperature (colder). Based on this principle, it is possible to correct for the tilt angle. Furthermore, it is also feasible to use more than one pair, for calculating not only the roll angle but also the pitch angle. However, even though this method is very useful for larger AVS, it has poor functionality for NAVs, especially during indoor missions where this kind of IR tracking works poorly because the IR radiation signals are more or less unpredictable.

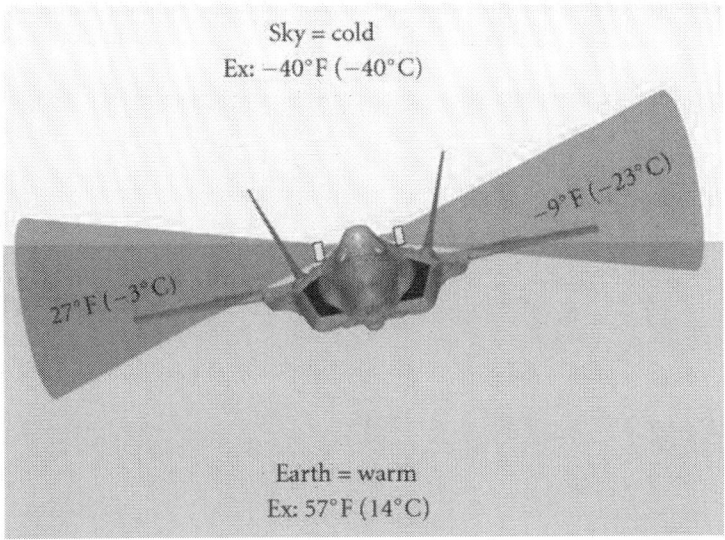

Figure 23. Example of angle compensation using two IR sensors reproduced from [64] under GNU FDL 1.2.

For NAV applications, in which each extra sensor means additive space, power, and weight, the system should be kept as simple as possible, using for example very low drift gyroscopes or accelerometers or using some compensation circuitry.

The other class of sensors is the data-collecting sensors that provide useful information for the users. Examples are cameras, microphones, gas sensors, biological sensors, radiation sensors, and so forth. Depending on the applications, most AVS will include one or more of these data sensors. Cameras and microphones are two of the most useful data sensors for AVS. A camera is required to help the user pilot the vehicles when there is no direct vision between them (e.g., in indoor missions). Similarly as for batteries, smart phones and other consumer electronics have driven the development. The smallest available cameras nowadays are based on CMOS sensors which offer "advantages in on-chip functionality, system power reduction, low cost and miniaturization" [65] when compared with the earlier used CCD image sensors.

Microphones are useful in spying, rescue operations, and similar applications. Today, micro- and nanotechnologies allow building very small microphones, such as those in [66] with diaphragm dimension of

2.1×2.1mm² (Figure 24) using a "single-crystalline wafer as the substrate for the microphone capsule" [66].

Figure 24. Example of ultrasmall microphone, diaphragm [66].

Micro- and nanotechnologies also provide great improvements for gas sensors, since "the sensitivity of chemical gas sensors is strongly affected by the specific surface of sensing" [67]. Using nanotechnologies, it is possible to build nano structures with a larger sensing area and either keeping the sensitivity constant while reducing the size or increasing the sensitivity while keeping the size constant. Many different materials can be used for building gas sensors. Some examples can be found in [68] where metal oxide is used, or in [69] that presents gas sensors based on conducting polymers. Depending on the application, a range of other sensors can also be used. In this case, important selection criteria for the choice of sensor are small dimensions, low weight, and low power consumption.

Actuators are needed for different applications on board AVS. They are used for flight control, for instance, making the vehicle turn, for moving the sensors, for example, movable cameras or for building useful tools, such as micropliers for picking up samples. Similarly as for flapping-wing systems, linear actuators are theoretically the most suitable solution for this application. Although there are a lot of studies of new materials and new concepts for linear actuators, all the existing prototypes have limited maximum elongation and/or long response time that limit the applicability on board AVS. The suboptimal solution is using micro servo actuators that are rotary actuators. They consist of a small electric motor, with some cogwheels that form a micro gear. In 2002, a micro harmonic drive was realized as one of the smallest micro backlash-free servo actuator in the world [70] with a size of 6 mm in diameter (another version is available with 8 mm diameter size) and with only 1 mm axial length. It is made with nickel-iron, and it has an output torque of 50 mNm (for the 8 mm diameter gear). Other rotary actuator technologies, such as piezoelectric motors [71] or shape alloy motors [72], are still under development, and they are yet not mature enough for AVS applications. Table 4 shows a quality comparison of some linear and rotary actuators.

Table 4. Quality comparison of actuators: adapted from [35, 79]

Actuators	Advantages	Disadvantages
Linear actuators		
Piezoelectric Ceramic	(i) Excellent performances except strain output (ii) Strain output can be magnified using bender arrangements	(i) Require high activation voltage
Shape memory alloy	(i) Excellent performance except frequency range	(i) Poor fatigue life
Magnetostrictor	(i) Excellent performances except strain output	(i) Require high activation voltage
Solenoid	(i) High strain	(i) Low energy density
Electroactive Polymers (EAP)	(i) Dielectric elastomers outperform muscle in both stress and strain output.	(i) Only ionic EAPs operate on low voltage (ii) Novel technology not widely available
Rotary actuators		
Electric motors	(i) Efficiency (ii) Reliability (iii) Versatility	(i) Weight (ii) Dimensions

FUTURE TRENDS

Considering the short-term future (1–3 years from mid-2010), rotary-wing NAV will be the most important commercial type, since it has the best performing technology at present and the near future. Prox Dynamics [17] has prototypes that will be available on the market in 2011. They present a system consisting of two parts: the NAV and the ground control station (GCS). The GCS (shown in Figure 25) has three different uses: (1) tt will protect the vehicles during the transport, (2) it will be used as remote control system for the NAV in flight. (3) As a station for recharging the AVS batteries. The whole system (NAV + GCS) will have a total weight less than 1 kg, and dimensions of 15 cm × 15 cm × 5 cm [2] (Figure 26).

Machine Component Design

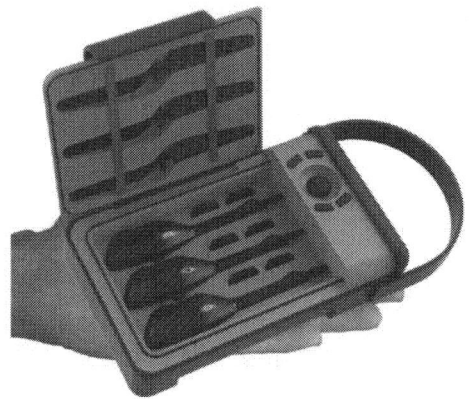

Figure 25. Graphic representation of the GCS from Prox Dynamics, containing three NAVs in the bottom side while the controls and the LED screen are placed on the upper side [17].

Figure 26. The Hornet 2-b (Prox Dynamics), complete with camera and video transmitter [17].

These systems will be customizable by the clients and equipped with different sensors depending on mission needs.

Specifications for those AVS state that will be able to fly for up to 30 minutes with a 10 m/s maximum speed. It has a video camera and transmitter for the video signal to the GCS.

However, in the future, flapping-wing solutions are viable and will improve maneuverability and efficiency relative to rotary NAVs. Several technologies can potentially replace batteries as power supplies in future NAV. Recent advances make ultra-capacitors a good candidate. However, depending on future developments, fuel cells are also promising, in particular direct methanol fuel cells, where the fuel storage is less complicated than for hydrogen-based fuel cells.

Further size and weight reductions of communication systems are important issues for the future. Micro- and nanoelectromechanical systems (MNEMS) technologies can be used to provide devices, such as lighter, smaller, and less power consuming resonators and filters than the current state of-the-art devices. However, reducing antenna dimensions while keeping acceptable performance is challenging since antenna performance is related to size governed by the laws of electromagnetic radiation. Furthermore, the transmission power cannot be reduced under a certain threshold without degrading the quality of the communication. MNEMS actuators will also replace the relative heavy rotary actuators (based on small electric motors), with lighter and more energy efficient linear actuators based on new materials, such as electro active polymers. However, the improvements will be relatively small, since the avionics on board of an NAV are already ultraminiaturized (a few grams).

Future NAVs will most likely be equipped with GPS and radar systems. Infrared and/or high-definition cameras could be included. Mission-specific sensors and actuators could be removed and replaced depending on the application. A quick-connection method can help to, for example, rapidly replace a gas sensor with a radiation sensor in a few seconds. Nanotechnology could play an important role also in aerodynamic improvements. For instance, a combination of different thin deposited layers of functional materials could allow the realization of morphable wings. These wings will be able to change their shape in accordance with the flying regime in order to maximize, in real-time, the efficiency of the vehicle. The shape changing can for example be on the attach angle or on the wing surface roughness.

Future trends could also include the development of sophisticated software that will enable operating future ultra-small NAVs in coordinated swarms. Furthermore, with the future improvements of artificial intelligence, some of them will have decision-making capabilities, opening the way to completely new mission profiles.

CONCLUSIONS

In this article, the main typologies of air vehicle systems used for micro- and nano-air vehicles have been presented. The typologies are fixed-wing, rotary-wing, flapping-wing and passive AVS. For each type, the main features are described, with a particular focus on their advantages and disadvantages. Several examples of existing prototypes have been described, and a final comparison between fixed-wing, rotary-wing, flapping-wing, and passive AVS have been presented. The rotary-wing principle is at present, and in the near future, in most NAV applications, the best option since they are capable to hover and have good maneuverability in NAV range dimensions. Military surveillance and reconnaissance are the most promising applications for such vehicles. Flapping-wing typology adapted from nature, first of all entomopters adapted from insects, is a promising future option, but more long-term research is needed to make this typology practical for AVS in general and for NAVs.

Power requirements and power sources for NAVS are major challenges for present and future NAVs, resulting in limited flight times around or less than 30 minutes and payload capabilities around 10 grams or less. Rechargeable batteries, mostly rechargeable lithium ion batteries will in the near future, remain the main power source in NAVs, with most of the power used by the electric motors. Fuel cells and ultra-capacitors show promising potentials to be used in the future, but more research and development are needed to make them practical and with high enough power density for use in NAVs. In addition, the development of a new class of gas turbines, which have already been proven promising, can surprise us in the near future. Sensors will play an increasingly important role, as they will be more and more used for improving flight control and collecting various information data from the environment during missions. For special missions and requirements, new sensor classes have to be developed; they will be focused on having low weight and low power consumption. Communications will remain a challenge as the power consumption scales poorly with size reduction of the NAVs. Micro- and nanotechnologies are enabling technologies for NAVs that will evolutionarily contribute to further size and weight reductions, improved sensors and actuators for better flight control and data collection during missions, and improved energy sources with higher power densities.

ACKNOWLEDGMENTS

This work is performed as part of a contract research for Prox Dynamics in the "Mosquito" project, supported by the Norwegian Research Council. Trygve Fredrik Marton and Geir Morten Mellem from Prox Dynamics are acknowledged for their support and suggestions for this paper.

REFERENCES

1. C. Galiński and R. Zbikowski, "Some problems of micro air vehicles development," Bulletin of the Polish Academy of Sciences: Technical Sciences, vol. 55, no. 1, pp. 91–98, 2007.
2. D. H. Paulsen, "Nano UAS- an upcoming reality," in Proceedings of the 24th International Unmanned Air Vehicles Conference, Bristol, UK, 2009.
3. R. J. Bachmann, "Biologically inspired mechanisms facilitating multi modal locomotion for areal micro-robot," in Proceedings of the 24th International Unmanned Air Vehicles Conference, Bristol, UK, 2009.
4. S. V. Serokhvostov, "Ways and technologies required for MAV miniaturization," in Proceedings of the European Micro Air Vehicle Conference (EMAV '08), Braunschweig, Germany, July 2008.
5. W. Shyy, Y. Lian, J. Tang et al., "Computational aerodynamics of low Reynolds number plunging, pitching and flexible wings for MAV applications," Acta Mechanica Sinica, vol. 24, no. 4, pp. 351–373, 2008.
6. M. James and C. M. S. F. McMichael, "Micro Air Vehicles—Toward a New Dimension in Flight," 1997, http://www.fas.org/irp/program/collect/docs/mav_auvsi.htm.
7. T. Hylton, "Nano Air Vehicle program," 2010, http://www.darpa.mil/dso/thrusts/materials/multfunmat/nav/index.htm.
8. D. J. Pines and F. Bohorquez, "Challenges facing future micro-air-vehicle development," Journal of Aircraft, vol. 43, no. 2, pp. 290–305, 2006.
9. U. Yearbook, UAS: The Global Perspective, vol. 164, UAS Yearbook, 7th edition, 2009/2010.
10. R. J. Wood, S. Avadhanula, E. Steltz et al., "An autonomous palm-sized gliding micro air vehicle—design, fabrication, and results of a fully integrated centimeter-scale MAV," IEEE Robotics and Automation Magazine, vol. 14, no. 2, pp. 82–91, 2007.
11. S. V. Serokhvostov, "Flapping wings efficiency investigation on the basis of physical law conservation," in Proceedings of the European Micro Air Vehicle Conference (EMAV '08), Braunschweig, Germany, 2008.

12. I. Kro, F. Prinz, and M. Shantz, "A miniature rotorcraft concept phase II final report," Tech. Rep., Stanford University, Palo Alto, Calif, USA, 2001.
13. I. Kroo and P. Kunz, "Development of the mesicopter a miniature autonomous rotorcraft," inProceedings of the American Helicopter Society Vertical Lift Aircraft Design Conference, American Helicopter Society, San Francisco, Calif, USA, 2000.
14. S. Bouabdallah, P. Murrieri, and R. Siegwart, "Towards autonomous indoor micro VTOL," Autonomous Robots, vol. 18, no. 2, pp. 171–183, 2005.
15. S. Bouabdallah, M. Becker, and R. Siegwart, "Autonomous miniature flying robots: coming soon!," IEEE Robotics and Automation Magazine, vol. 14, no. 3, pp. 88–98, 2007.
16. H. Wu, D. Sun, and Z. Zhou, "Micro air vehicle: configuration, analysis, fabrication, and test,"IEEE/ASME Transactions on Mechatronics, vol. 9, no. 1, pp. 108–117, 2004.
17. Prox Dynamics, http://www.proxdynamics.com.
18. J. Happel and H. Brenner, Low Reynolds Number Hydrodynamics: With Special Applications to Particulate, Springer, New York, NY, USA, 1983.
19. T. J. Mueller, "Aerodynamic measurements at low raynolds numbers for fixed wing micro-air vehicles," Tech. Rep., University of Notre Dame, Notre Dame, The Netherlands, 2000.
20. I. M. A.-Q. A. A. M. Al-Bahi, "Micro aerial vehicles design challenges: state of the art review," inProceedings of the SAS UAV Scientific Meeting & Exhibition, Jeddah, Saudi Arabia, 2006.
21. Z. L. A. J.-M. Moschetta, "Rotary vs. flapping-wing nano air vehicles: comparing performances," inProceedings of the European Micro Air Vehicle Conference (EMAV '09), Delft, The Netherlands, 2009.
22. M. Wing and J. F. Gieras, "Calculation of the steady state performance for small commutator permanent magnet DC motors: classical and finite element approaches," IEEE Transactions on Magnetics, vol. 28, no. 5, pp. 2067–2071, 1992.
23. P. Ferrat, C. Gimkiewicz, S. Neukom, Y. Zha, A. Brenzikofer, and Thomas Baechler, "Ultra-miniature omni-directional camera for an autonomous flying micro-robot," in Optical and Digital Image Processing, Proceedings of SPIE, April 2008.
24. J. M. Grasmeyer, "Development of the black widow micro air vehicle," in Proceedings of the 39th AIAA Aerospace Sciences Meeting and Exhibit, M. T. Keennon, Ed., American Institute of Aeronautics and Astronautics, 2000.
25. F. Zhang, R. Zhu, P. Liu, W. Xiong, X. Liu, and Z. Zhou, "A novel Micro Air Vehicle with flexible wing integrated with on-board electronic devices," in Proceedings of the IEEE International Conference on Robotics, Automation and Mechatronics (RAM '08), pp. 252–257, September 2008.

REFERENCES

26. D. Schafroth, S. Bouabdallah, C. Bermes, and R. Siegwart, "From the test benches to the first prototype of the muFly micro helicopter," Journal of Intelligent and Robotic Systems, vol. 54, no. 1–3, pp. 245–260, 2009.
27. A. Datta, "The martian autonomous rotary-wing vehicle (MARV)," Tech. Rep., University of Maryland, College Park, Md, USA, 2000.
28. A. Datta, B. Roget, D. Griffiths et al., "Design of a Martian autonomous rotary-wing vehicle," Journal of Aircraft, vol. 40, no. 3, pp. 461–472, 2003.
29. World's Lightest Micro-Flying Robot Built by Epson, Available from: http://www.physorg.com/pdf860.pdf.
30. G. B. Kim, "Design and performance test of rotary wing micro air vehicle without tail rotor," in Proceedings of the European Micro Air Vehicle Conference (EMAV '08), Braunschweig, Germany, July 2008.
31. Corporation, "A. T-Rex 250," http://www.align.com.tw/html.
32. J. Kim, M. S. Kang, and S. Park, "Accurate modeling and robust hovering control for a quad-rotor VTOL aircraft," Journal of Intelligent and Robotic Systems, vol. 57, no. 1–4, pp. 9–26, 2010.
33. G. R. Spedding and P. B. S. Lissaman, "Technical aspects of microscale flight systems," Journal of Avian Biology, vol. 29, no. 4, pp. 458–468, 1998.
34. R. Madangopal, Z. A. Khan, and S. K. Agrawal, "Energetics-based design of small flapping-wing micro air vehicles," IEEE/ASME Transactions on Mechatronics, vol. 11, no. 4, pp. 433–438, 2006.
35. A. Conn, S. Burgess, R. Hyde, and C. S. Ling, "From natural flyers to the mechanical realization of a flapping wing micro air vehicle," in Proceedings of the IEEE International Conference on Robotics and Biomimetics (ROBIO '06), pp. 439–444, December 2006.
36. P. Brochu and Q. Pei, "Advances in dielectric elastomers for actuators and artificial muscles,"Macromolecular Rapid Communications, vol. 31, no. 1, pp. 10–36, 2010.
37. Y. Fujihara, T. Hanamoto, and F. Dai, "Fundamental research on polymer material as artificial muscle," Artificial Life and Robotics, vol. 12, no. 1-2, pp. 232–235, 2008.
38. M. A. A. Fenelon and T. Furukawa, "Design of an active flapping wing mechanism and a micro aerial vehicle using a rotary actuator," Mechanism and Machine Theory, vol. 45, no. 2, pp. 137–146, 2010.
39. H. Tanaka and I. Shimoyama, "Forward flight of swallowtail butterfly with simple flapping motion,"Bioinspiration and Biomimetics, vol. 5, no. 2, Article ID 026003, 2010.
40. H. Tanaka and R. J. Wood, "Fabrication of corrugated artificial insect wings using laser micromachined molds," Journal of Micromechanics and Microengineering, vol. 20, no. 7, Article ID 075008, 2010.

41. R. J. Wood, S. Avadhanula, E. Steltz et al., "Design, fabrication and initial results of a 2g autonomous glider," in Proceedings of the 31st Annual Conference of IEEE Industrial Electronics Society (IECON '05), pp. 1870–1877, November 2005.
42. J. Kuhle, an Airship for Cooperative Robot Tasks, IFAC, Portugal, 2003.
43. J. Saunders and R. Beard, "Tracking a target in wind using a micro air vehicle with a fixed angle camera," in Proceedings of the American Control Conference (ACC '08), pp. 3863–3868, June 2008.
44. U. Pesavento and Z. J. Wang, "Flapping wing flight can save aerodynamic power compared to steady flight," Physical Review Letters, vol. 103, no. 11, Article ID 118102, 2009.
45. M. Cavendish, Growing Up with Science, Marshall Cavendish, Tarrytown, NY, USA, 3rd edition, 2006.
46. T. T. H. Ng and G. S. B. Leng, "Design optimization of rotary-wing micro air vehicles," Proceedings of the Institution of Mechanical Engineers, Part C: Journal of Mechanical Engineering Science, vol. 220, no. 6, pp. 865–873, 2006.
47. S. J. Morris, "Design and flight test results for micro-sized fixed-wing and VTOL aircraft," inProceedings of the International Conference on Emerging Technologies and Factory Automation, 1997.
48. I. Sher, D. Levinzon-Sher, and E. Sher, "Miniaturization limitations of HCCI internal combustion engines," Applied Thermal Engineering, vol. 29, no. 2-3, pp. 400–411, 2009.
49. A. Mehra, X. Zhang, A. A. Ayon, I. A. Waitz, M. A. Schmidt, and C. M. Spadaccini, "Six-wafer combustion system for a silicon micro gas turbine engine," Journal of Microelectromechanical Systems, vol. 9, no. 4, pp. 517–527, 2000.
50. Onera Company, http://www.onera.fr/defa/micro-machines-thermiques-mems.
51. X. C. Shan, Z. F. Wang, R. Maeda, Y. F. Sun, M. Wu, and J. S. Hua, "A silicon-based micro gas turbine engine for power generation," in Proceedings of the Symposium on Design, Test, Integration and Packaging of MEMS/MOEMS (DTIP '06), Stresa, Italy, April 2006.
52. R. R. Glassock, et al., "Multimodal hybrid powerplant for unmanned aerial systems (UAS) robotics," inProceedings of the 24th Bristol International Unmanned Air Vehicle Systems Conference, University of Bristol, Bristol, UK, 2009.
53. D. C. Müller, Energy Storage: Strong Momentum in High-End Batteries, Credit Suisse, Suisse, 2007.

REFERENCES

54. S. K. Mandal, P. S. Bhojwani, S. P. Mohanty, and R. N. Mahapatra, "IntellBatt: towards smarter battery design," in Proceedings of the 45th Design Automation Conference (DAC '08), pp. 872–877, June 2008.
55. Fuel Cell Store, http://www.fuelcellstore.com.
56. DARPA, "Micro air vehicle powered entirely by fuel cell makes debut flight," in News, April 2003.
57. IEEE Spectrum, "Circuit Could Swap Ultra capacitors for Batteries," June 2001, http://spectrum.ieee.org/semiconductors/design/circuit-could-swap-ultracapacitors-for-batteries.
58. Maryland, U.o. 2010, http://www.nanocenter.umd.edu/ news/news_story.php?id=3773.
59. B. Kim, R. N. Candler, M. Hopcroft, M. Agarwal, W. T. Park, and T. W. Kenny, "Frequency stability of wafer-scale encapsulated MEMS resonators," in Proceedings of the13th International Conference on Solid-State Sensors and Actuators and Microsystems (TRANSDUCERS '05), pp. 1965–1968, Stanford University, June 2005.
60. Discera, "Mems based oscillators," http://www.discera.com.
61. K. M. Strohm, F. J. Schmückle, B. Schauwecker, J. F. Luy, and W. Heinrich, "Silicon micromachined RF MEMS resonators," in Proceedings of the IEEE MSS-S International Microwave Symposium Digest, pp. 1209–1212, June 2002.
62. S. A. Bhave, DI. Gao, R. Maboudian, and R. T. Howe, "Fully-differential poly-SiC lamé-mode resonator and checkerboard filter," in Proceedings of the 18th IEEE International Conference on Micro Electro Mechanical Systems (MEMS '05), pp. 223–226, February 2005.
63. M. Palaniappan, et al., "Unmanned Vehicle Group Semester Report," Tech. Rep., Purdue University Aeronautics and Astronautics, 2010.
64. The Paparazzi Project, http://paparazzi.enac.fr/wiki/Main_Page.
65. L. Q. Wang, Y. Shi, Z. Lu, and H. Duan, "Miniaturized CMOS imaging module with real-time DSP technology for endoscope and laryngoscope applications," Journal of Signal Processing Systems, vol. 54, no. 1–3, pp. 7–13, 2009.
66. T. Tajima, T. Nishiguchi, S. Chiba et al., "High-performance ultra-small single crystalline silicon microphone of an integrated structure," Microelectronic Engineering, vol. 67-68, pp. 508–519, 2003.
67. B. Ding, M. Wang, J. Yu, and G. Sun, "Gas sensors based on electrospun nanofibers," Sensors, vol. 9, no. 3, pp. 1609–1624, 2009.
68. S. Palzer, E. Moretton, F. H. Ramirez, A. Romano-Rodriguez, and J. Wöllenstein, "Nano- and microsized metal oxide thin film gas sensors," Microsystem Technologies, vol. 14, no. 4-5, pp. 645–651, 2008.

69. H. Bai and G. Shi, "Gas sensors based on conducting polymers," Sensors, vol. 7, no. 3, pp. 267–307, 2007.
70. MicroHarmonicDeviceGroup, "Hollow shaft micro servo actuators realized with the Micro Harmonic Drive," 2002, http://www.mikroharmonicdrive.de.
71. J. Liu, Z. Yang, H. Zhao, and G. Cheng, "Novel precision piezoelectric step rotary actuator," Frontiers of Mechanical Engineering in China, vol. 2, no. 3, pp. 356–360, 2007.
72. I. Spinella, G. Scirè Mammano, and E. Dragoni, "Conceptual design and simulation of a compact shape memory actuator for rotary motion," Journal of Materials Engineering and Performance, vol. 18, no. 5-6, pp. 638–648, 2009.
73. A. Santhana and K. J. D. Jacob, "Effect of regular surface perturbations on flow over an airfoil," inProceedings of the 35th Fluid Dynamics Conference, American Institute of Aeronautics and Astronautics, Toronto, Canada, 2005.
74. CITIZEN MICRO CO., LTD., 2009, http://www.citizen-micro.com/tec/corelessmotor. html.
75. M. Bronz, et al., "Towards a long endurance MAV," International Journal of Micro Air Vehicles, vol. 1, no. 4, pp. 241–254, 2009.
76. G. Elert, http://hypertextbook.com/facts/2005/MichelleFung.shtml.
77. Technologies, M., http://www.maxwell.com.
78. Institute, G.E.N, "Fuel cell," http://www.geni.org/globalenergy/library/technical-articles/generation/title-page-images/fuelcell.jpg.
79. C. Tetrault, http://www.rctoys.com/rc-products-catalog/RC-AIRPLANES. html.

CITATION

Luca Petricca, Per Ohlckers, and Christopher Grinde, "Micro- and Nano-Air Vehicles: State of the Art," International Journal of Aerospace Engineering, vol. 2011, Article ID 214549, 17 pages, 2011. doi:10.1155/2011/214549.

CHAPTER 5

Radial Ball Bearings with Angular Contact in Machine Tools

Lubomír Šooš1

[1] STU Bratislava, Institute of manufacturing systems, environmental technology and quality management, Bratislava, Slovakia

INTRODUCTION

The decisive criteria of the quality of machining tools are their productivity and working accuracy.

One innovated method for improving the technological parameters of manufacturing machines (machine tools) is to optimise the structure of their nodal points and machine components.

Because of the demands on machine tool productivity and accuracy, the spindle-housing system is the heart of the machine tool, Figure 1, [1]. Radial ball bearings with angular contact are employed in ever increasing arrays. The number of headstocks supported on ball bearings with angular contact is increasing proportionally with the increasing demands on the quality of the machine tool [2]. This is because these bearings can be arranged in various combinations to create bearing arrangements which can enable the reduction of both radial and axial loads. The possibility of varying the number of bearings, their preload value, dimensions and the contact angle of bearings used in the bearing nodes, creates a broad spectrum of combinations which enable us to achieve the adequate stiffness and high speed capabilities of the **Spindle-Bearings System** (SBS) [2], [3]. Adequate stiffness and revolving speed of the headstock are necessary conditions for meeting the manufacturing precision quality and machine tool productivity required by industry.

When designing a machine tool headstock, the starting point is the design of the spindle support, as this limits the stability, accuracy and production capacity of the machine by its stiffness and revolving speed. However, the parameters influencing the stiffness and frequency can *act in opposition to each other*. The selection of the type of bearing has to take into consideration the optimization of its stiffness and revolving speed characteristics. The maximum turning speed of the bearings is a function of the maximum revolving speed of the individual bearings, their number, pre-load magnitude, manufacturing precision, and the types of lubrication used.

The stiffness of the SBS depends on the stiffness of the bearings and the spindle itself. There are several methods that can be employed for determining the static stiffness of the spindle system, e.g. [1] and [2].

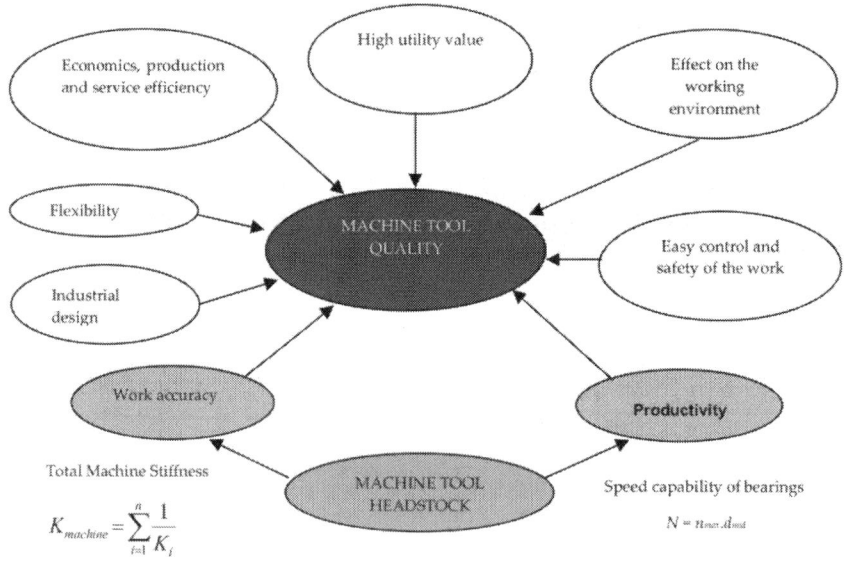

Figure 1. Factors influencing the Quality of Machine Tools, [4].

However, one problem which has not yet been solved is the calculation of the stiffness of the bearings, (or nodes of bearings) in the individual housing. Accurate calculation of the stiffness of the bearing nodes requires the determination of the static parameters of each bearing. From a mathematical point of view, this can be solved by using a system of non-linear differential equations, which requires the use of computers. To simplify the design, we need a static analysis which provides the basis for the dynamic characteristics of the mounting, and of the machine itself. Designers often prefer the conventional and proven methods of mounting, without taking into account the technical and technological parameters of the machine.

For the design engineer, it is important to be able to undertake a quick evaluation of various SBS variants at the preliminary design stage. The success of the design will depend on the correct choice of suitable criteria for the SBS, and if the design engineer has adequate experience in this field.

Headstock–The Heart of the Machine Tool

The headstock, whether tool or workpiece carrier, has a direct influence on the static and dynamic properties of the cutting process, Figure 2. The spindle-bearing system (SBS) stiffness affects the surface quality, profile, and dimensional accuracy of the parts produced. It also has a direct influence on machine tool productivity because the width of cut influences the initiation of self-induced vibration;it is directly proportional to machine tool stiffness and damping.

Complex analysis of the SBS is very difficult and complicated, [5]. The analysis requires an advanced understanding of mathematics, mechanics, machine parts, elasto-hydrodynamic theory, rolling housing techniques, and also programming skills. The results of our research into SBS have been divided into three parts:

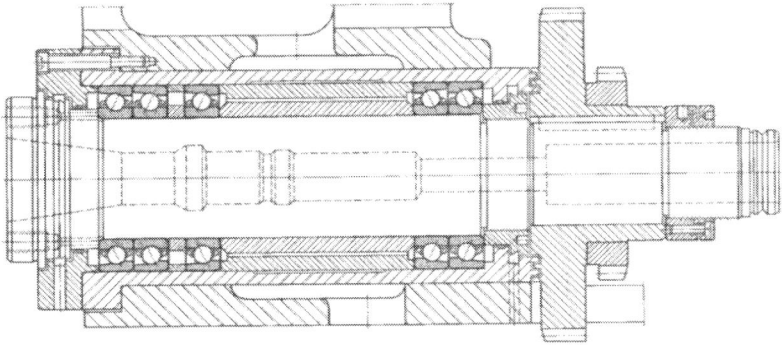

Figure 2. CNC profile milling machine, Carl Hurt Maschines, Germany; Work nodal point – 3x7013ACGA/P4, Opposite nodal point – 2x7013ACGA/P4

- Theoretical research - dealing with creating mathematical models
- Experimental research - verifying theoretical hypotheses and results on testing devices
- Application research - dealing with the special software application, *Spindle Bearings* for SBS design.

THEORETICAL RESEARCH

A modular structure of the theoretical research is shown in Figure 3.

Primary Static Analysis
Speed
The productivity of a machine tool, (Figure 1) can be increased in at least two different ways:

1. Externally - by shortening working time - within a working cycle
2. Internally - by reducing machining times (increasing the cutting width) - technological issues.

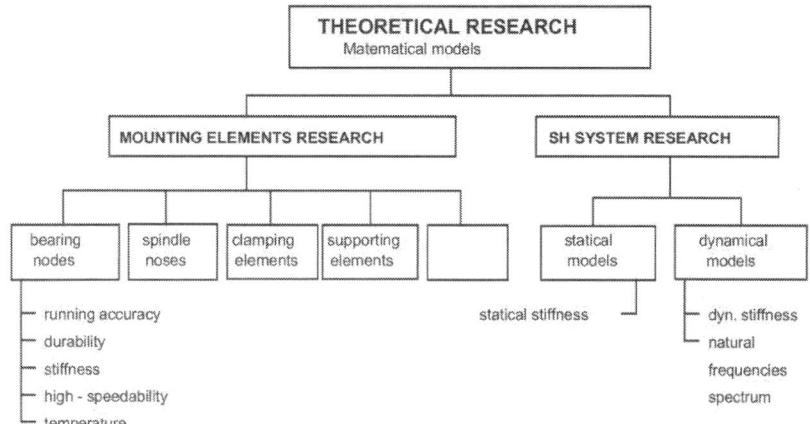

Figure 3. Modular structure of theoretical research, [5]

THEORETICAL RESEARCH

The philosophy of intelligent manufacturing systems applied to production processes minimise lost time. Further reducing lost time is expensive and has limited effectiveness at current levels of technological development. It has been shown that increased productivity can be achieved for example by changing the cutting speed. However this has a direct effect on tool life and on the dynamic stability of the cutting process.

The cutting speeds in machining processes depend on the technology applied, the cutting tool, and the workpiece material. The cutting speed also relates directly to the high-speed capability, and average diameter, of the bearings, the so-called factor $N = n_{max}.\ d_{mid}.$. Thus, from the point of view of the required cutting speed, the most important factor is the revolving frequency capacity of a spindle which is supported on a bearing system.

The calculation of the headstock's maximum revolving speed is relatively simple. The highest revolving speed of a bearing node is calculated on the basis of the highest revolving speed of one bearing, multiplied by various coefficients reflecting the influence on the bearings, the bearing arrangement, bearing precision, their preloaded value, and lubrication and cooling conditions.

Stiffness

The total static stiffness of machine tools is, in almost all cases, limited by the stiffness of the weakest parts. Amongst all the elements, the Spindle-Bearings System of the machine tool plays the most important role.

From results of structural analyses, the headstock can be considered as the heart of the whole machine tool. The design and quality of the machine tool must respect the quality of the drives and their features.

The headstock (as tool, or work piece carrier), has a direct influence on the static and dynamic properties of the cutting process. The Spindle-Bearing System's stiffness also influences the final surface quality, profile, and dimensional accuracy of the work piece.

The problem here is how quickly the headstock stiffness can be calculated with sufficient precision. The headstock stiffness must be calculated according to the deflection at the front end of the spindle, because the deflection at this point directly affects the precision of the finished product. The deflection at the spindle front end is the accumulation of various other, more or less important, partial distortions. The radial headstock stiffness can be calculated as follows:

$$K_{rc} = \frac{F_r}{y_{rc}}$$

(1)

The individual headstock parts, (spindle and bearing arrangement), create a serial spring arrangement and it is evident that the resulting stiffness Krc is limited by the stiffness of the weakest part. An expert can see which part should be improved, and which partial distortions need to be minimized.

Simplified Method of Calculation

The calculation of spindle front end deflection, which takes into consideration all the important parameters, can only be achieved by using powerful computers. The analysis can be carried out by standard or custom software programs.

Calculating the many combinations of SBS arrangements is very demanding on time and money. Undertaking stiffness analysis using standard programs depends on the engineers' experience. The results can be open to questionable even when a suitable mathematical method is used (finite element method, boundary element method, Castilian's theorem, graphic Mohr's method, etc). This is because the headstock box, bearings or bearings nodes are statically indefinite systems which produce a nonlinear deformation of the node when under load.

Special software programmes are very expensive. They are developed using the most up-to-date theoretical and practical knowledge. These programs have been developed by research institutions and bearing producers and the possibility of using such programs significantly influences their position on the SBS market. Taking the above into account, engineers would benefit from the existence of a simplified method of static analysis. Such a methodology would enable the engineer at the preliminary design stage to limit the number of possible spindle-bearing variants and determine the direction which would lead to the optimal SBS design, [6].

The main methodological advantage of computer analysis is the possibility of repeating single calculating algorithms in a matrix shape. To this end a special software package, "Spindle Headstock" was developed at the Department of Production Engineering in the Faculty of Mechanical Engineering at STU, Bratislava, [7].

THEORETICAL RESEARCH

The resulting radial deformation, *yrc*, of the front spindle end is shown in Figure 4.

Resulting static distortion of the front-end spindle equals

$$y_{rc} = y_0 + y_1 + y_t + y_a + y_v + y_{sb} + y_h \tag{2}$$

Our experience has shown that whatever mathematical method and software is used, the spindle distortion caused by *bending moments y0* a nd by *bearing compliances yl* have the greatest influence on the resulting front end spindle distortion, [6].

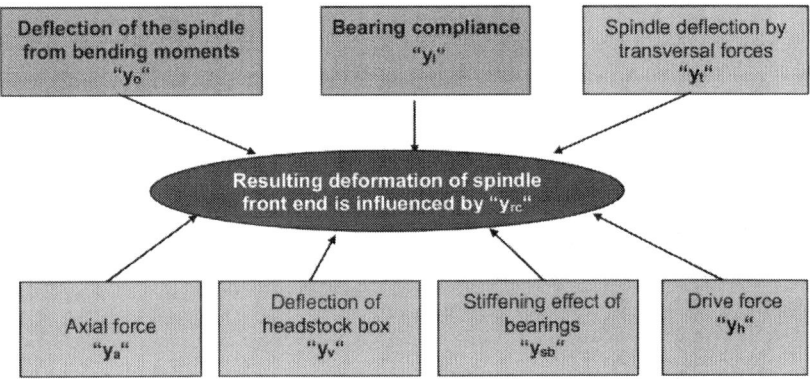

Figure 4. Factors influencing the resulting deflection

Then

$$y_{rc} = y_0 + y_1 \tag{3}$$

where the distortion caused by bending moments is as follows:

$$y_0 = \frac{F_r a^2}{3E} \left[\frac{a}{J_a} + \frac{L}{J_L} \right] \tag{4}$$

and the deflection caused by bearing compliance is as follows:

$$y_l = \frac{F_r}{L^2}\left[\frac{a^2}{K_B} + \frac{(L+a)^2}{K_A}\right]$$

(5)

Increasing moments of inertia "J_a", "J_L" were calculated as follows:

$$J_a = \frac{\pi}{64}\left[D_a^4 - d_a^4\right] \text{ and } J_L = \frac{\pi}{64}\left[D_L^4 - d_L^4\right]$$

(6)

The definition of the quantities is shown in Figure 5. The individual headstock parts (spindle, bearing arrangement,) create a serial spring arrangement, and it is evident that the resulting stiffness Krc is limited by the stiffness of the weakest part, [1] [2] and [9].

At the same time, parameters "F_r", "a", "L" influence the value of both deflections. The spindle deflection caused by bending moments can be decreased by the following methods:

- increasing the material modulus of elasticity "E",
- increasing moments of inertia "J_a", "J_L" by a change of spindle diameters "D_a", "D_L", "d_a", "d_L".

The resulting static distortion of the spindle front-end can be explicitly described by a multi-parametrical equation in the form of:

$$y_F = f\,[E,\, F_r,\, a,\, L,\, J_{a,},\, (D_a,\, d_a)J_L\,(D_L,\, d_L)\,,\, K_A,\, K_{B,}\rho]$$

(7)

- spindle material and dimensions (E, Da, da, DL, dL)
- loading forces position, orientation and magnitude (Fr, N, rF, b)
- bearing arrangement configuration and stiffness (KA, KB,)
- spindle and bearing arrangement space configuration (L, a)
- spindle box construction (kξ, ρ)

There remains one significant problem with the calculation of the bearing nodes, and that isthat they are statically indeterminate systems.

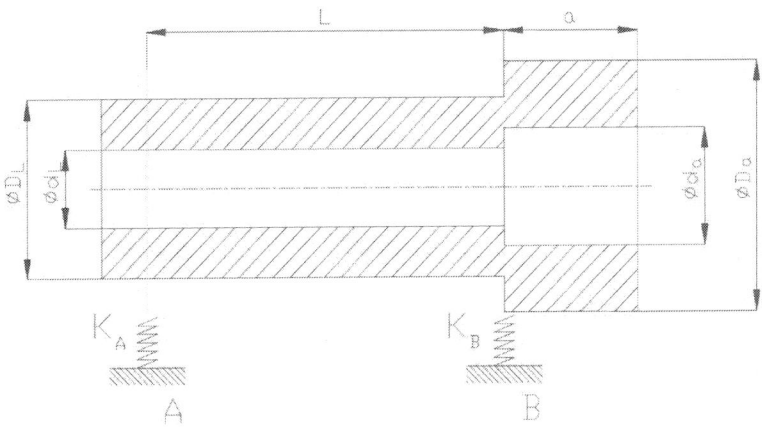

Figure 5. Cross section scheme of the spindle-bearing system

Dynamic. Analysis

While static analysis of SBS describes spindle behaviour in a static mode, dynamic analysis describes SBS behaviour under real conditions, in real running time, and so the real operational state is better represented. It is very important to know the dynamic characteristics, especially in high-speed headstocks. It is important to ensure that the operational revolving frequencies do not fall within the resonant zone. When this happens, the vibration amplitude of the spindle is considerably increased, and the spindle's total stiffness falls to unacceptable levels.

The most common determining dynamic characteristics of SBS are:

- the spectrum of natural (resonant) frequencies (usually the first three frequencies),
- the amplitudes of vibrations along the spindle independent of the revolving frequencies of the spindle,
- the resonant amplitudes of vibrations,
- the dynamic stiffness of the spindle (at the given speed of the spindle).

The SBS dynamic properties (dynamic deflection of spindle front-end, natural frequencies spectrum) [5], are affected by factors shown in Figure 6.

Mathematical models for determining the dynamic properties of a spindle

Currently, the only reliable method for determining dynamic properties is to use experimental measurements. Therefore it is very useful to create reliable mathematical models for determining these dynamic properties.

In line with spindle mass reduction, mathematical models are divided into:

1/ discrete with 1°, 2° and N° degrees of freedom,
2/ continuous.

The discrete mathematical model developed for measuring the revolving vibration of spindles with N° degrees of freedom is worked out in [1], [5]. This mathematical model for calculating the dynamic properties of the spindle enables us to include in our calculation the effects of the materials, the dimensions of the rotating parts, the bearing node stiffness, and the radial forces generated by the cutting process and drive. The results calculated reflect a spectrum of natural frequencies and the dynamic deflection of the spindle under discrete masses.

The deflection of spindle yiloaded with concentrated forces at the ith point can be expressed in the form:

$$y_i = a_{i1}F_{1o} + a_{i2}F_{2o} + \ldots + a_{ik}F_{ko} + \ldots + a_{in}F_{no} \quad (m)$$

(8)

where aik(m/N) is Maxwell's affecting factor. Every mass point on the spindle produces centrifugal force

$$F_{io} = m_i y_i \omega^2 \quad (N)$$

(9)

where mi(kg) is mass i^{th} discrete segment.

THEORETICAL RESEARCH

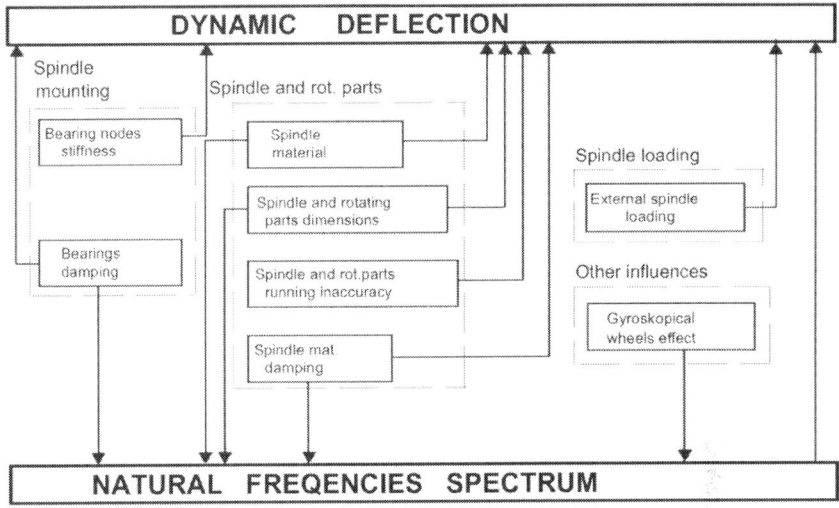

Figure 6. Factors affected by SBS dynamic properties

The application of the aforementioned equations and their modification for masses of "n" value, will create a system of homogenised algebraic equations where the results of determinant D are angular natural frequencies of the transverse vibrations of the spindle $\omega_i(rad.s^{-1})$.

The procedure for determining the dynamic deflections y_i, when the dimensions of the spindle, rotating parts, stiffness of bearing arrangement, and the external radial forces are taken into consideration, is very similar to the previous one. These procedures are described in [6].

$$\Delta = \begin{vmatrix} 1 - a11\ m1\ \omega^2 & - a12\ m2\ \omega^2 & \ldots & - a1n\ m\ n\ \omega^2 \\ - a21\ m1\ \omega^2 & 1 - a22\ m2\ \omega^2 & \ldots & - a2n\ m\ n\ \omega^2 \\ \ldots & \ldots & \ldots & \ldots \\ - a\ n1\ m1\ \omega^2 & - a\ n2\ m2\ \omega^2 & \ldots & 1 - a\ nn\ m\ n\ \omega^2 \end{vmatrix} = 0$$

(10)

It is relatively easy to transform this mathematical model into a computer readable format and the calculation of the dynamic characteristics can be quickly achieved.

Theoretical Research on Bearing Nodes
Arrangements Of Nodal Points
Usually, radial ball bearings with angular contact arrangements in their nodal points contain 2, 3 or more bearings, see Figure 7.

Criteria For Selecting The Arrangement Of Bearings
The number of spindle bearing systems supported on ball bearings with angular contactincreases proportionally with increasing demand on the machine tool. By varying the bearings and their arrangement in the bearing nodes (DB, DF, DT, TBT,TTF, QBC,..), the value of the contact angle, magnitude of preload, and type of flanges can be optimized to suit the required, resulting stiffness and speed-capability of the spindle-bearing system.

In order to assess the maximum permissible speed of different typesof spindle rotations, a parameter for the so-called high-speed characteristics has been introduced: N = n*max*.d*mid*, where "n*max*" denotes the maximum spindle revolutions and "d*mid*"the medial diameter of the bearings. Following this parameter, roller bearings of machine tool spindles can be divided into 3 basic groups, [10]:

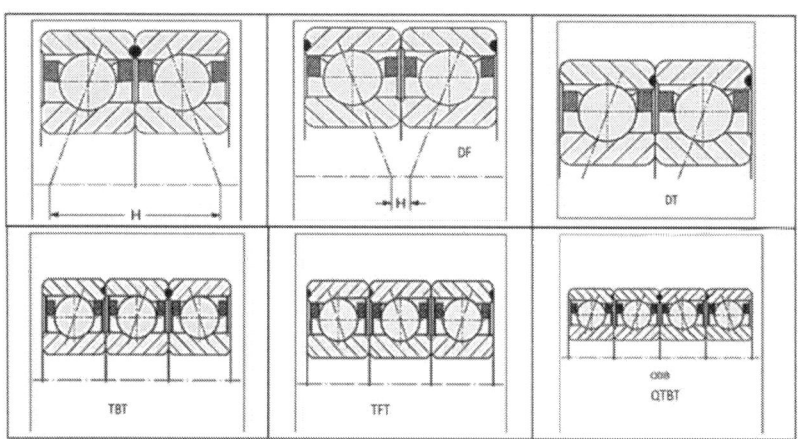

Figure 7. Arrangements in nodal points

GROUP 1: $N = (0,1 - 0,5).10^6$: Headstocks of heavy duty machines for turning, milling and drilling operations. In these machines the spindles are predominantly mounted using double-row roller bearings in combination with axial ball bearings, or tapered bearings. We can assume that the linearization of the deformation curve in roller bearings is sufficiently accurate, which simplifies the calculation of the radial stiffness of the nodal point, [2]. These *mountings* offer high stiffness and load-bearing capacity and quiet operation.

GROUP 2: $N = (0.4 - 1).10^6$ is characterized by bearings of medium size and are found in smaller NC and CNC turning, milling, drilling and grinding machine tools. The maximum possible speed in bearings with linear joints is limited by the heat produced in the head, and therefore they are used only for the mounting of spindles with the lowest values of coefficient N. Developments in the field of increased speed capability is focused on bearings with point contacts, as these have better friction characteristics.

GROUP 3: Spindles mounted in bevelled radial bearings with optimized structure (design) and using new composite materials enabling high-speed operation, $N = (0.8 - 2.5) \times 10^6$, which is typical for high-speed machining.

Spindle mountings using only radial bevelled bearings, (table 1), [11] can be divided into 2 basic types:

- spindles mounted on bearing nodes with "directionally" arranged bearings, with equal orientation of contact angles in each nodal point 1, 2, 3, and 7, table 1.
- spindles mounted on nodal points with bearings arranged according to shape. Bearings are arranged in "O" or (X) shape, in combination with "T".

A typical feature of the nodes of spindle bearings is the application of pre-stressing, which provides the stiffness of the nodal point and reduces any skidding of the rollers at high revolutions.

Pre-stressing can be achieved through three flange design principles:

a. Sprung flange: thermal expansion (dilatation) is eliminated by changing the length of the elastic materials positioned between the flange and the bearings, which ensures minimum change in the pre-stress value.
b. Stiff (Rigid) flange: provided by a fixing nut or casing. This design provides better stiffness characteristics. The pre-stress value is changed due to the influence of thermal dilatation.
c. Controllable flange: axially adjustable (by means of hydraulics), which ensures the required pre-stressing for different operational conditions.

The highest values of the coefficient N can be achieved by using spindles mounted on nodes with a "directional" arrangement of bearings, 1, 2 and 3. When used in conjunction with the controllable flange, the correct types of lubrication and cooling, speeds which are comparable with the maximum revolutions of the bearings themselves can be achieved. Thus they can be applied in high-speed machining [11]. These mounting types, in combination with the sprung support, are mostly used for grinding.

For difficult technological operations requiring considerably higher stiffness in the radial and axial directions, nodal points with bearings arranged according to shape, together with fixed supports are typical.

There is negligible use of hybrids of the basic types of mounting (mounting 5), as shown in table 1. In such cases one nodal point has bearings arranged according to shape, while the other has directionally arranged bearings, (Figure 2). The pre-stressing in the front nodal point is ensured by a stiff flange, and in the rear nodal point by a sprung flange.

Table 1. Type of SBS using radial ball bearings with angular contact [11]

Seq. No.	CONFIGURATION		$N = n_{max}.d_{mid}.10^6$ [mm.min⁻¹]	Characteristic	Use
	Rear bearing node	Forward bearing node			
1.	$t_1=1, t_2=0$	$t_1=0, t_2=1$	1,2 - 2,5	- single direction of rotation - light axial and radial loads	- grinding internal holes
2.	$t_1=1, t_2=0$	$t_1=0, t_2=2$	0,8 - 1,6	- suitable for extremely short spindles - medium axial loads	- finishing machines - drilling of deep holes
3.	$t_1=2, t_2=0$	$t_1=0, t_2=2$	0,8 - 1,4	- medium radial loads - very common method of use	- grinding internal holes - milling - drilling
4.	$t_1=1, t_2=1$	$t_1=1, t_2=1$	0,6 - 1	- machining light metals - medium radial loads	- grinding - precision drilling - turning/lathe
5.	$t_1=1, t_2=0$	$t_1=1, t_2=2$	0,5 - 0,9	- medium axial loads	- drilling of deep holes - milling
6.	$t_1=1, t_2=1$	$t_1=1, t_2=2$	0,4 - 0,9	- medium axial loads - very common method of use	- turning/lathe - drilling
7.	$t_1=2, t_2=0$	$t_1=3, t_2=0$	0,3 - 0,6	- high axial loads medium radial loads	- milling - boring

Stiffness

The stiffness of the bearing arrangement (KA, KB) is the specific parameter which influencestheconsequent spindle distortion. Wehave developed a simplified mathematical model for calculating radial and axial stiffness, [11], [12].

THE CALCULATION OF RADIAL STIFFNESS OF NODAL POINTS

Assumptions Of Solution

According to the Hertz assumptions [13], [14], there is a dependence between load "P" and deformation "δ" at the contact point of the ball with the plane, given by the relationship

$$P = k_\delta \cdot \delta^{3/2}$$

(11)

a. the bearings in the nodal points are of the same type and dimensions, with precise geometric dimensions
b. the value of the contact angle is the same for all directionally-arranged bearings in the nodal point, which delivers equal distribution of strain on these bearings
c. radial load is equally distributed onto all the bearings in the nodal point

Stiffness of Nodal Points with Directionally-Arranged Bearings

The calculation of the stiffness of a nodal point is based on the stiffness of the bearing itself [15], which is defined as:

$$K_{r1} = \frac{dF_{r1}}{d\delta_{r0}}$$

(12)

As radial displacement δ_{r0} is a function of contact deformation δ_0 of the ball with the highest load [13], the equation for calculating the stiffness of bevelled radial bearings will have the form:

$$K_{r1} = \frac{dF_{r1}}{d\delta_0} \cdot \frac{d\delta_0}{d\delta_{r0}}$$

(13)

When calculating stiffness, the distribution of load among the rollers must be determined, and the dependence between the load on the top ball and external load must be found. The distribution of load in the bearing can be derived from the balance under static conditions [14],

$$F_{r1} = \frac{F_{r1}}{i} = \sum_{j=0}^{z} P_j \cdot \cos(\alpha_j) \cdot \cos(j \cdot \gamma)$$

(14)

where

$\gamma = 360/z$ is the spacing angle of the balls.

The values of contact deformations δj and angles αj differ from each other around the circumference of the bearing and can be expressed as follows, (Figure 8).

$$\delta_j = l_{rj} - l_p = \sqrt{[l_z \cdot \sin(\alpha_z) + \delta_p]^2 + [l_z \cdot \cos(\alpha_z) + \delta_{r0} \cdot \cos(j \cdot \gamma)]^2} - l_p$$

(15)

$$\cos(\alpha_j) = \frac{l_z \cdot \cos(\alpha_z) + \delta_{r0} \cdot \cos(j \cdot \gamma)}{\sqrt{[l_z \cdot \sin(\alpha_z) + \delta_p]^2 + [l_z \cdot \cos(\alpha_z) + \delta_{r0} \cdot \cos(j \cdot \gamma)]^2}}$$

(16)

By loading the pre-stressed bearing with a radial force, the distance, $O_A O_{ip}$, between the centre of the balls is constant, (Figure 8 b, c).

$$l_p \cdot \sin(\alpha_p) = l_{rj} \cdot \sin(\alpha_{rj}) = konst.$$

(17)

The dependence between the deformation of the jth ball and the top ball can be determined by the relation

$$\delta_j = \delta_0 \cdot \cos(j \cdot \gamma)$$

(18)

By derivation of equation (14) we get

$$\frac{dF_{r1}}{d\delta_0} = i \cdot \sum_{j=0}^{z} \left[\frac{dP_j}{d\delta_j} \cdot \cos(\alpha_j) - P_j \cdot \sin(\alpha_j) \cdot \frac{d\alpha_j}{d\delta_j} \right] \cdot \frac{d\alpha_j}{d\delta_0} \cdot \cos(j \cdot \gamma)$$

(19)

The unknown derivatives in equation (19) can be calculated by changing the relations (11), (17), (18).

$$\frac{dPj}{d\delta j} = \frac{3}{2} k_\delta^{2/3} P_j^{1/3}$$

(20)

$$\frac{d\alpha_j}{d\delta_j} = -\frac{tg(\alpha_j)}{l_{rj}}$$

(21)

$$\frac{d\delta_j}{d\delta_0} = \cos(j \cdot \gamma)$$

(22)

The interdependence of the contact deformation and radial displacement, Figure 8, can be determined from the relation

THE CALCULATION OF RADIAL STIFFNESS OF NODAL POINTS

$$\frac{d\delta_0}{d\delta_{r0}} = \left(\frac{d\delta_j}{d\delta_0}\right)^{-1} \cdot \frac{d\delta_j}{d\delta_{r0}}$$

(23)

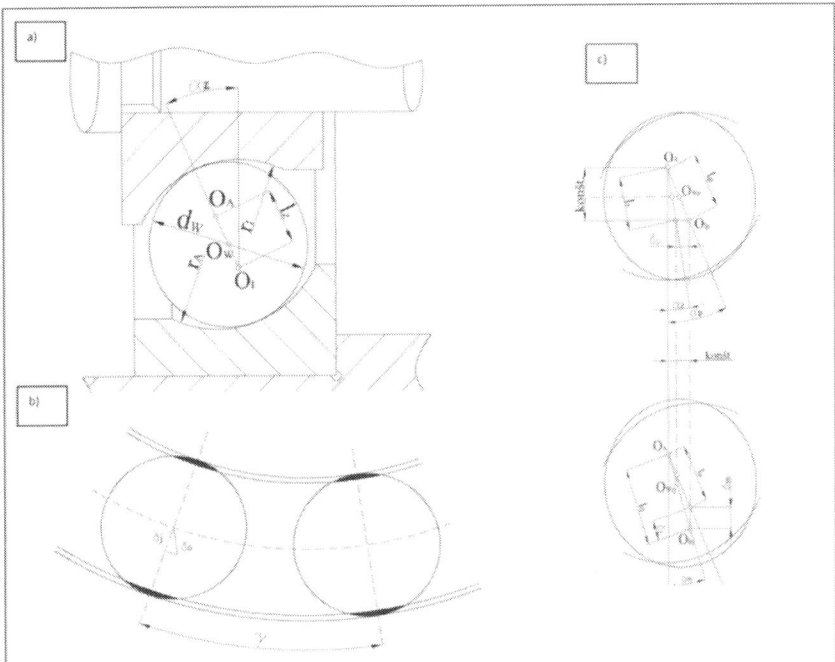

Figure 8. Detailed bearing scheme, a – unloaded, b – pre-stressed, c – radial loaded

Where $d\delta_j/d\delta_{r0}$ is calculated from equation (15)

$$\frac{d\delta_j}{d\delta_{ro}} = \frac{1}{2} \cdot \frac{2\left(1_z \cos\alpha_z + \delta_{ro}\cos(j.\gamma)\right)\cos(j.\gamma)}{\sqrt{\left(1_z \cos\alpha_z + \delta_{ro}\cos(j.\gamma)\right)^2 + \left(1_z \sin\alpha_z + \delta_p\right)^2}} = \cos\alpha_j \cos(j.\gamma)$$

(24)

by inserting equations (24) and (22) into equation (23)

$$\frac{d\delta_0}{d\delta_{r0}} = \frac{1}{\cos(j.\gamma)} \cdot \cos(\alpha_j) \cdot \cos(j.\gamma) = \cos(\alpha_j)$$

(25)

After inserting equations (25) and (19) into equation (13) we get the resulting relation for the stiffness of a pre-stressed nodal point with directionally-arranged bearings.

$$K_r = i \cdot \sum_{j=0}^{z} \left[\frac{3}{2} \cdot k_\delta^{2/3} \cdot P_j^{1/3} \cdot \cos^2(\alpha_j) + P_j \cdot \frac{\sin^2(\alpha_j)}{l_{rj}} \right] \cdot \cos^2(j.\gamma)$$

(26)

Stiffness of Nodal Point with Bearings Arranged According To Shape
When calculating the nodal point with bearings arranged according to shape, we divide the nodal point into part "1" and part "2" (Table 1), with the *same* orientation of contact angles in nodes with directionally-arranged bearings, and the stiffness of the parts is calculated as follows:

$$K_{r1} = i_1 \cdot \sum_{j=0}^{z} \left[\frac{3}{2} \cdot k_\delta^{2/3} \cdot P_j^{1/3} \cdot \cos^2(\alpha_{1j}) + P_j \cdot \frac{\sin^2(\alpha_{1j})}{l_{r1j}} \right] \cdot \cos^2(j.\gamma) \quad (a)$$

$$K_{r2} = i_2 \cdot \sum_{j=0}^{z} \left[\frac{3}{2} \cdot k_\delta^{2/3} \cdot P_j^{1/3} \cdot \cos^2(\alpha_{2j}) + P_j \cdot \frac{\sin^2(\alpha_{2j})}{l_{r2j}} \right] \cdot \cos^2(j.\gamma) \quad (b)$$

(27)

For example in Figure 9 the total numbers of bearings in the front node SBS is 5: $i1 = 3$, $i2 = 2$, contact angles $\alpha 1 = \alpha 2 = 25$.

We determine the total stiffness of the nodal point by the addition of both parts of the node with the equation:

$$K_r = K_{r1} + K_{r2}$$

(28)

In order to optimize the stiffness and load-bearing capacity for specified technological conditions, the manufacturers of machine tools have come out with a new, non-traditional solution for nodal points. By diminishing the contact angle of the bearing in Part 2, the axial stiffness of the nodal point is partially decreased, but at the same time, the value of the radial stiffness and boundary axial load is increased.

Approximate Calculation of Stiffness

When evaluating the overall stiffness of a spindle, the designer must take into account the approximate calculation of the stiffness of the nodal points.

If all the balls are loaded, and there are more than 2 per bearing [14], the following equation can be applied:

$$\sum_{j=0}^{z} \cos^2(j \cdot \gamma) = \frac{z}{2}$$

(29)

If the bearing angle is loaded only in an axial direction by the pre-stressing force, then the load on the rollers is constant around the whole circumference and can be expressed, for the particular parts of the nodal point [11] in the form

$$P_{1j} = \frac{F_p}{i_1 \cdot z \cdot \sin(\alpha_{p1})}; P_{2j} = \frac{F_p}{i_2 \cdot z \cdot \sin(\alpha_{p2})}$$

(30)

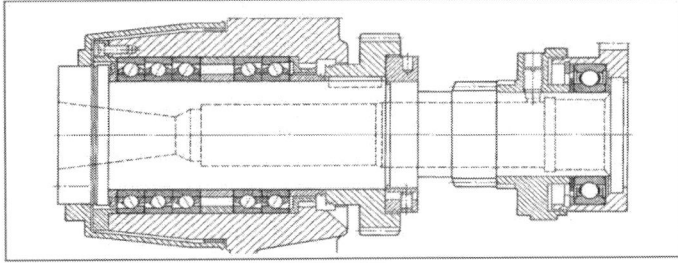

Figure 9. Horizontal machining centre, Thyssen-HüllerHille GmbH, Germany;Work nodal – 3x71914 ACGB/P4 - 2x71914 ACGB/P4, Opposite side– 6011-2Z

If the magnitude of the spindle bearing contact angles is not greater than 26 degrees, then the value of the second expression in equations (27a) and (27b) is negligible.

Taking these assumptions into consideration, we obtain the relationship for the approximate calculation of the radial stiffness of a bearing angle with directionally placed bearings in the form:

$$K_r = \frac{3 \cdot 10^{-3}}{4} \cdot z^{2/3} \cdot k_\delta^{2/3} \cdot i^{2/3} \cdot F_p^{1/3} \cdot \frac{\cos^2(\alpha)}{\sin^{1/3}(\alpha)}$$

(31)

and with bearings arranged according to shape in the form:

$$K_r = \frac{3 \cdot 10^{-3}}{4} \cdot z^{2/3} \cdot k_\delta^{2/3} \cdot i_1^{2/3} \cdot F_p^{1/3} \cdot \frac{\cos^2(\alpha_1)}{\sin^{1/3}(\alpha_1)} \cdot \left[1 + \frac{i_2^{2/3} \cdot \cos^2(\alpha_2) \cdot \sin^{1/3}(\alpha_1)}{i_1^{2/3} \cdot \cos^2(\alpha_1) \cdot \sin^{1/3}(\alpha_2)}\right]$$

(32)

where the approximate value of the deformation constant is

$$c_\delta = 10^5 \cdot \sqrt{1,25 \cdot d_w}$$

(33)

d_w – is the diameter of the balls.

The pre-stressing value "F_p" can be calculated according to the standard, STN 02 46 15. Some foreign manufacturers (for example, SKF, FAG, SNFA...) publish this value in their catalogues. The number of balls "z" and their diameters "d_w" of some types of bearings are quoted in the literature, e.g. [16].

- Based on the equation for the calculation of a nodal point axial stiffness [17]

THE CALCULATION OF RADIAL STIFFNESS OF NODAL POINTS

$$K_a = \frac{3.10^{-3}}{2} z^{\frac{2}{3}} \cdot k_\delta^{\frac{2}{3}} \cdot i_1^{\frac{2}{3}} \cdot F_p^{\frac{1}{3}} \cdot \sin^{\frac{5}{3}} \alpha_1 \left[1 + \frac{i_2^{\frac{2}{3}} \cdot \sin^{\frac{5}{3}} \alpha_1}{i_1^{\frac{2}{3}} \cdot \sin^{\frac{5}{3}} \alpha_2} \right]$$

(34)

and substituting the equation in brackets

$$T_1 = 1 + \frac{i_2^{\frac{2}{3}} \cos^2 \alpha_2 \sin^{1/3} \alpha_1}{i_1^{\frac{2}{3}} \cos^2 \alpha_1 \sin^{1/3} \alpha_2} \text{ in equation}$$

(35)

and

$$T_2 = 1 + \frac{i2^{\frac{2}{3}} \sin^{\frac{5}{3}} \alpha_2}{i1^{\frac{2}{3}} \sin^{\frac{5}{3}} \alpha_1} \text{ in equation}$$

(36)

the dependence between the axial and radial stiffness can be expressed by the relation

$$K_r = \frac{K_a}{2} \cdot \frac{1}{tg2\alpha_1} \cdot \frac{T_2}{T_1}$$

(37)

When α1 = α2

in a nodal point with bearings arranged according to shape, or i = 0 in nodal points with bearings arranged according to direction, the quotient of the constants T1, T2 will be equal to 1 and the relation (37) will be simplified. Thus

$$K_r = \frac{K_a}{2} \cdot \frac{1}{tg2\alpha}$$

(38)

Taking equations (32) and (34) into consideration, it is evident that the stiffness of the bearing arrangement depends on the number of bearings (i1 and i2) in the arrangement, the dimensions of the bearings (z1, dw1 and z2, dw2), the contact angle (α1 and α2) and the preload value Fp.

Conclusions of the Analysis
The conclusions of the analysis [10] are as follows:

- Radial stiffness increases proportionally with increasing values of "z", "dw", "i", "Fp" and decreases when "a" is prolonged, (Figure 10).
- The parameters "z" and "dw" must be evaluated in mutual interaction because they characterize the size and dimensions of the bearings. Increasing both of these parameters, and producing the consequent increase of stiffness of the bearing arrangement, can be achieved by increasing the inner bearing diameter. The disadvantage here is that maximum revolving speed will be reduced. A more suitable solution is to decrease the width of the bearing, e.g. from B72 to B70, B719 or B718. In this case the number of rolling elements "z" will be increased and their diameter "dw" will be smaller.
- Considering equations (31), (32) and (34), it is evident that "z" has a more important influence on stiffness than "dw". If the diameter of the rolling elements is smaller, their weight will also be decreased, and this fact will allow an increase in the maximum revolving speed.
- The number of bearings in bearing arrangement "i" is the significant factor which can favourably influence stiffness. But the increased number of bearings will reduce the maximum revolving frequency and therefore it is possible to use this solution only for low speed spindle-bearing systems.
- The preload has a relatively small effect on the stiffness of bearing arrangements. The preload real value also depends on the type of flange used. When fixed flanges are used the preload value can exceed the nominal value by several times. This will cause excessive preload values which produce heat and the bearing arrangement will break down much sooner than expected.
- The contact angle "α" has a significant influence on the variation of the stiffness of the bearing arrangement. When the value of the contact

angle is increased, the radial stiffness and maximum revolving speed of the bearing arrangement is also decreased. On the other hand, the axial stiffness of the bearing arrangement will be significantly increased.

OPTIMIZATION OF THE SPINDLE-BEARING SYSTEM IN RELATION TO TEMPERATURE

The optimisation of SHS with regard to temperature

In addition to the bearing arrangements, the temperature properties of the bearing supporting node have an increasingly greater significance on the high-speed capability of the bearing. The main goal of thissection is to show the SHS design under real operating conditions, taking into consideration the temperature-related behaviour of the spindle and bearing nodes.

The value of the changes in SHS temperature depends on the temperature gradient, the type of bearing arrangement (DB, DF, DT, ...), the contact angle of the bearing, and the distance between the bearings arranged in the node.

The stiffness of the given example was analysed using the application software "Spindle Headstock" [3], developed in our department.

The analysis identified the optimal stiffness, which was then applied to the headstock of the DB 24 fy. Ex-Cell-O GmbH., Eislinger precision boring machine, Figure 11, [18]

The headstock used for analysis had the following parameters:

Output power: P = 3 kW

Maximum speed: n_{cmax} = 5500 min-1

Shape inaccuracy at boring E_s≤ 1,5µm

Surface roughness: Et = 1...1,5µm

Bearings: FAG B 7016 C.TPA.P4.UL in "O" arrangement

Bearing lubrication: grease

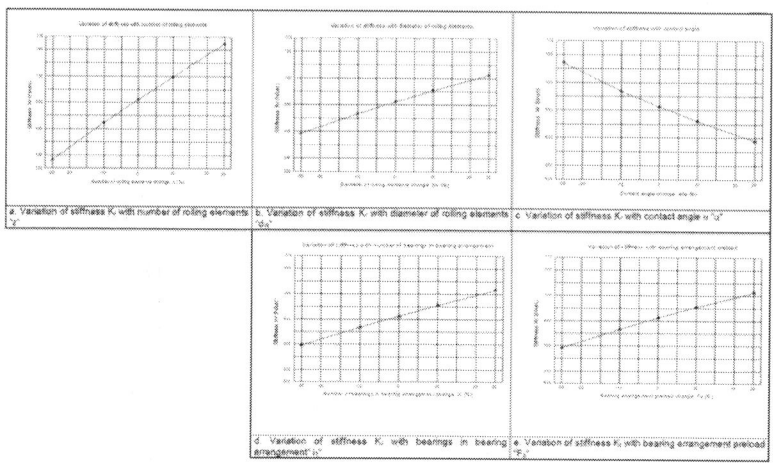

Figure 10. Variation of stiffness with bearing arrangement parameters

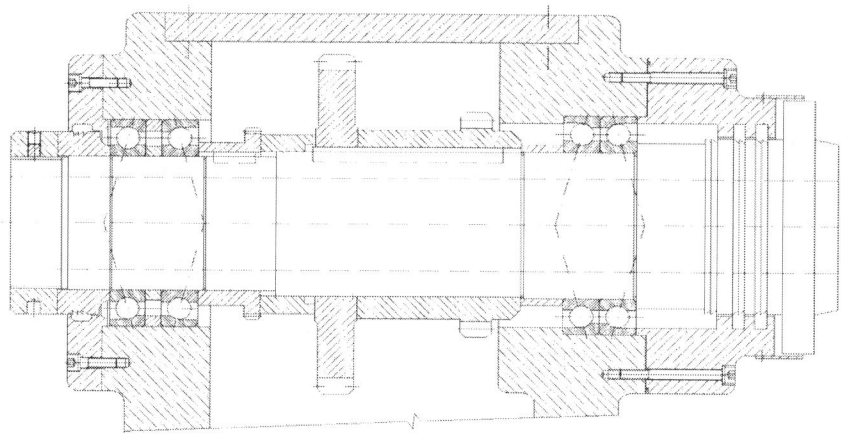

Figure 11. The Headstock of the precision boring machine DB 24 fy. Ex-Cell-O GmbH., Eislinger, [18]

OPTIMIZATION OF THE SPINDLE-BEARING SYSTEM IN RELATION TO TEMPERATURE

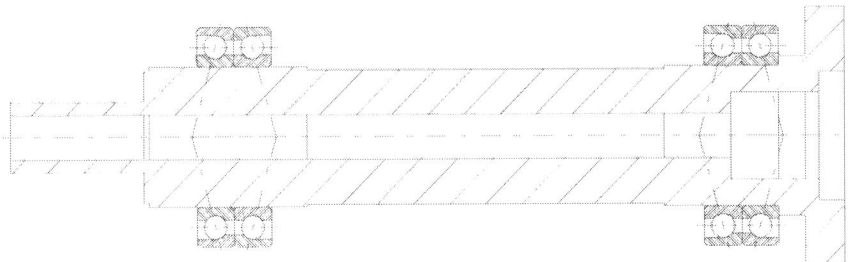

Figure 12. Model of the spindle

Results for spindle with arrangement DB - DB.

Radial load: $F_r = 1000$ N
Axial load: $F_a = 200$ N
Desired spindle speed: $n_{\alpha} = 5500$ min^{-1}
No driving force

Working conditions

Lubrication: Plastic grease
Cooling: Good cooling

	Rear support	Front support
Bearings		
- type:	2pc. [B 7016 CTB]	2pc. [B 7016 CTB]
- dimensions [mm]:	D=125 d=80 B=24 dW=13.49	D=125, d=80, B=24, dw=13.49
- arrangement:	◇	◇
- precision grade:	P4	P4
Preload:	Light	Light
Flange:	Fixed flange	Fixed flange
Maximum speed:	$Zn_{max} = 5\,256$ min^{-1}	$Zn_{max} = 5\,256$ min^{-1}
Pre-load:	$ZF_p = 404$ N	$PF_p = 402$ N
Reactions:	$R_A = 205$ N	$R_B = 1\,205$ N
Radial stiffness:	$K_{rA} = 666\,243$ Nmm^{-1}	$K_{rB} = 651\,216$ Nmm^{-1}
Axial stiffness:	$K_{aA} = 97\,860$ Nmm^{-1}	$K_{aA} = 97\,745$ Nmm^{-1}
Durability:	$T_{rvZ} = 394\,366$ hours	$T_{rvP} = 225\,577$ hours
Bearings distance:	$L = 297$ mm	
Total displacement at the end:	$y_{r(t+a)} = 0.00372931$ mm	
Total stiffness:	$K_{rc} = 268\,146$ Nmm^{-1}	
Optimal calculated values		
Optimal bearing length:	$L_{opt} = 283.6$ mm	
Optimal displacement at the end L_{opt}:	$y_{rmin} = 0.00372686$ mm	
Optimal stiffness:	$K_{rcopt} = 268\,322$ Nmm^{-1}	

The temperature dilatation of the spindle can be described by the equation:

$$\Delta L = \lambda_t \cdot L \cdot \Delta t$$

(39)

If the distance between the bearings in the "DB" arrangement is short (Figure 13a), the dilatations in a radial direction is greater, [18]. The temperature gradient causes the dilatation of the inner bearing rings to be greater than that of the outer rings. Consequently, the original preload increase in temperature will be higher in the bearing node. The elevated temperature will influence the temperature gradient, and the preload value could cause bearing node failure.

The preload change was defined by the change in the distance between the centres of the radii of the rolling raceways:

$$l_0 = r_A + r_I - d_w$$

(40)

The distance lt at given temperature gradient in accordance with Figure 13 is

OPTIMIZATION OF THE SPINDLE-BEARING SYSTEM IN RELATION TO TEMPERATURE

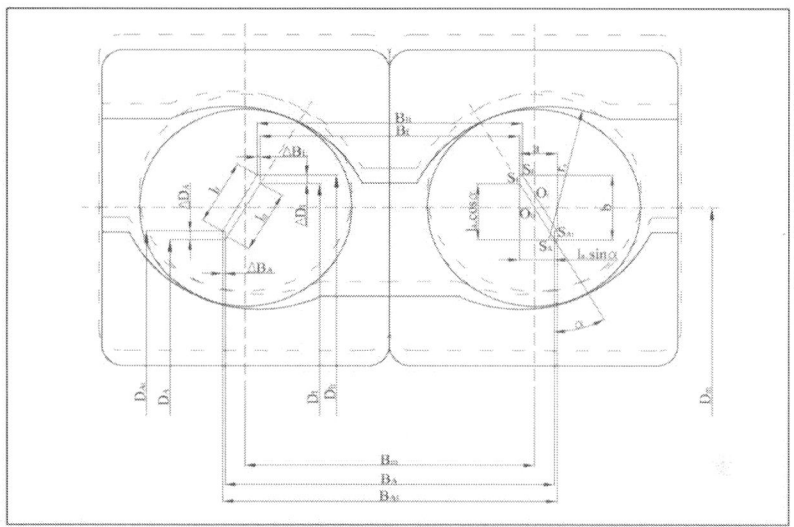

a) Distance between bearings $B_m = 29$ mm

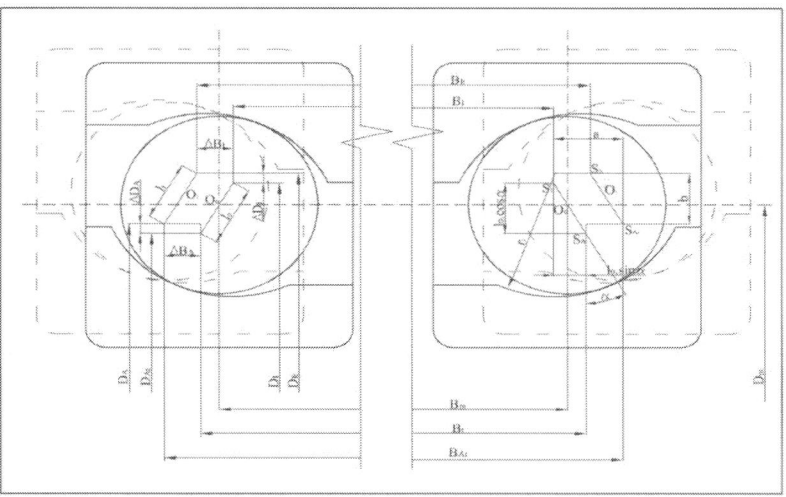

b) Distance between bearings $B_m = 500$ mm

Figure 13. Temperature deformation of bearing arrangement B 7016 C TPAP4UL in "DB"

$$l_t = \sqrt{a^2 + b^2}$$

(41)

Where

$$a = l_0 \cdot \sin\alpha + \frac{\lambda_t}{2} \cdot [B_m \cdot (t_I - t_A) + l_0 \cdot \sin\alpha \cdot (t_I + t_A - 2 \cdot t_0)]$$

(42)

$$b = l_0 \cdot \cos\alpha + \frac{\lambda_t}{2} \cdot [D_m \cdot (t_A - t_I) + l_0 \cdot \cos\alpha \cdot (t_I + t_A - 2 \cdot t_0)]$$

(43)

The magnitude of deformation will be

$$\Delta\delta = l_0 - l_t$$

(44)

and preload change in accordance with [18] will be

$$\Delta F = \Delta\delta \cdot z^{2/3} \cdot c_\delta^{2/3} \cdot \sin^{8/3}\alpha$$

(45)

where

$$c_\delta = 10^5 \cdot \sqrt{1,25 \cdot d_w}$$

(46)

In the twin bearings FAG B 7016 C.TPA.P4.UL with "DB" arrangement, at a temperature gradient of 10 °C, and with bearing distance Bm = 29 mm, the preload will increase by 13,32 N.

Conversely, if the distance of the bearing in "O" arrangement is long (Figure 13b), the dilatations in the axial direction prevail and cause a decrease in the value of the preload.

In the twin bearings FAG B 7016 C.TPA.P4.UL with "DB" arrangement, at a temperature gradient of 10°C, the preload will be decreased by 5,88 N.

OPTIMIZATION OF THE SPINDLE-BEARING SYSTEM IN RELATION TO TEMPERATURE

In "DB" arrangement, the main goal of temperature optimization is dependant on the determination of the optimum distance between the bearings at which a change in the preload at the given temperature gradient would be minimal.

In accordance with Figure 13 the condition

$$l_0 = l_t \tag{47}$$

must be satisfied.

By substituting equations (42) and (43) into (46), the optimal bearing separation distance from the point of view of temperature can be deduced from:

$$B_{mopt} = D_m \cdot \frac{\cos \alpha}{\sin \alpha} - \frac{l_0 \cdot (t_I + t_A - 2 \cdot t_0)}{t_I - t_A} \cdot \left(\frac{1}{\sin \alpha}\right) \tag{48}$$

Figure 14 shows the change of optimal bearing distance at various values of the temperature gradient for the analysed SBS, Figure 11.

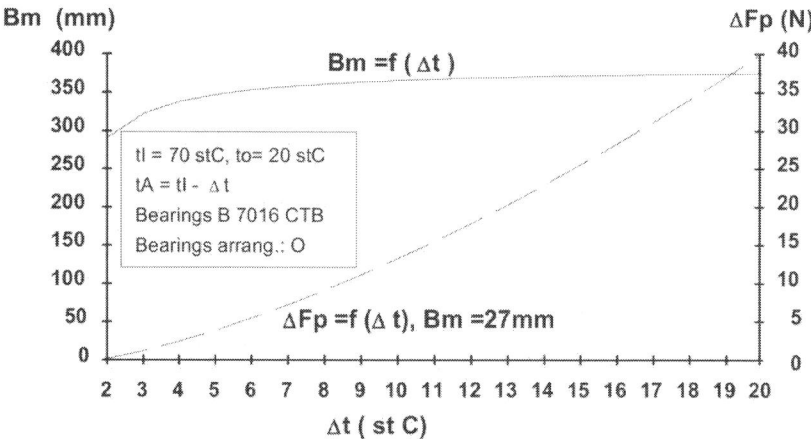

Figure 14. The inter - dependence of bearing preload change, ideal distance between bearings and change of temperature in the bearings arrangement system.

Recommendation for Improvements In Construction

The recommendations from the point of view of temperature optimisation for the DB 24 SHS boring machine are based on the results of the analysis undertaken. From the perspective of temperature, it can be seen that a change in bearing node arrangement to individual spindle supports from "DB" to "DT" would be advantageous, Figure 15.

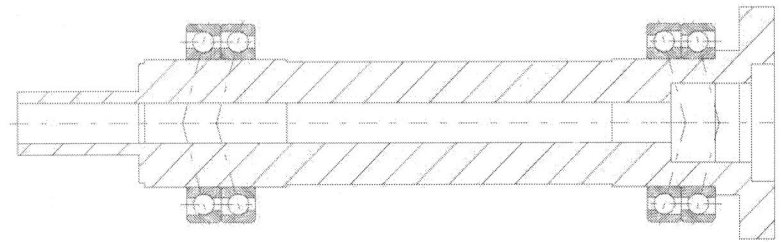

Figure 15. Model of the spindle

Results for spindle with arrangement DB - DB.

Radial load: $F_r = 1000$ N
Axial load: $F_a = 200$ N
Desired spindle speed: $n_{Ch} = 5500$ min^{-1}
No driving force

Working conditions

Lubrication: Plastic grease
Cooling: Good cooling

	Rear support	Front support
Bearings		
- type:	2pc. [B 7016 CTB]	2pc. [B 7016 CTB]
- dimensions [mm]:	D=125 d=80 B=24 dW=13.49	D=125, d=80, B=24, dw=13.49
- arrangement:	◇	◇
- precision grade:	P4	P4
Preload:	Light	Light
Flange:	Fixed flange	Fixed flange
Maximum speed:	$Zn_{max} = 5\ 256$ min^{-1}	$Zn_{max} = 5\ 256$ min^{-1}
Pre-load:	$ZF_p = 404$ N	$PF_p = 402$ N
Reactions:	$R_A = 205$ N	$R_B = 1\ 205$ N
Radial stiffness:	$K_{rA} = 666\ 243$ Nmm^{-1}	$K_{rB} = 651\ 216$ Nmm^{-1}
Axial stiffness:	$K_{aA} = 97\ 860$ Nmm^{-1}	$K_{aA} = 97\ 745$ Nmm^{-1}
Durability:	$T_{rvZ} = 394\ 366$ hours	$T_{rvP} = 225\ 577$ hours

Bearings distance: $L = 297$ mm
Total displacement at the end: $y_{r(L-a)} = 0.00372931$ mm
Total stiffness: $K_{rc} = 268\ 146$ Nmm^{-1}

Optimal calculated values

Optimal bearing length: $L_{opt} = 283.6$ mm
Optimal displacement at the end L_{opt}: $y_{rmin} = 0.00372686$ mm
Optimal stiffness: $K_{rcopt} = 268\ 322$ Nmm^{-1}

OPTIMIZATION OF THE SPINDLE-BEARING SYSTEM IN RELATION TO TEMPERATURE

In comparison with the original bearing node arrangement, the radial stiffness of the rearranged spindle-bearing system will drop slightly, but its axial stiffness will increase. The advantage of the reconFigured SBS is that at real mean values of temperature gradient, the SBS stiffness will be almost fixed.

Spindle Headstock

The application software is used for calculating the SBS of machine tools supported on rolling bearings. The programme enables us to determine all elements and calculate the properties of the spindles and shafts which are supported on rolling bearings. The application software enables very fast and user-friendly calculation of the radial spindle stiffness in the bearing arrangement in a bearing unit.

The architecture of the programme contains a number of mathematical formulae which have been experimentally verified. These models respect the conditions of the spindle working accuracy in terms of the external load cutting forces, driving forces, and also spindle rotation speed.

The basic interactive programme offers:

1. The ability to input user-determined conditions for the calculation and optimisation of the spindle fitting system (Figure 16);

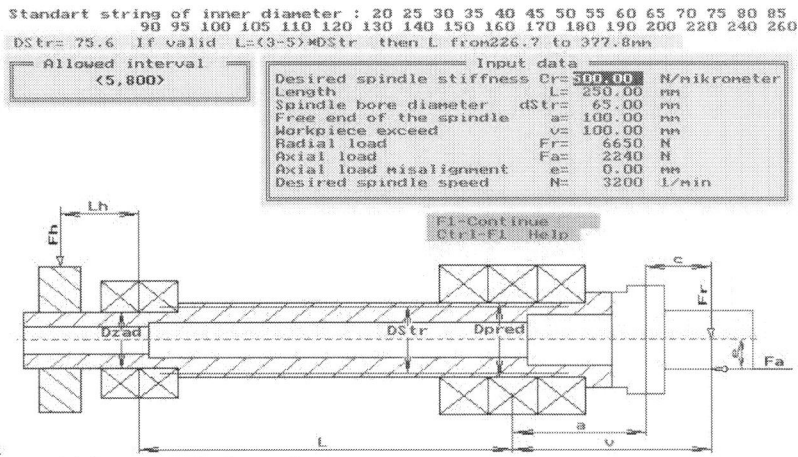

Figure 16. Entering the input data

1. The ability to select the most appropriate bearing or bearing node arrangements (Figure 17, Figure 18, Figure 19). Data about selected bearings can be ganed from extensive databases according to the users requirements within the bearing inner diameter range:
2. angular contact ball bearings, type 710..150 mm
3. single row cylindrical roller bearings, type N50..120 mm
4. full cylindrical roller bearings, type NN30..440 mm
5. axial angular contact bar, type 234425..380 mm
6. thrust ball bearings, single direction, type 5110..360 mm
7. thrust ball bearings, double direction, type 5210..190 mm
8. deep groove ball bearings, type 63..360 mm;

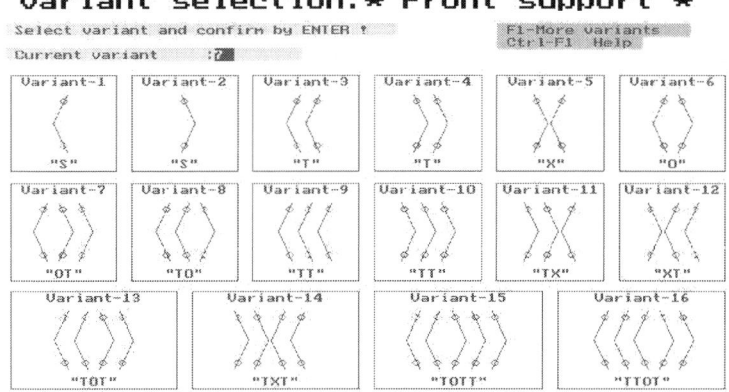

Figure 17. Selection of the bearing arrangement and type of bearings

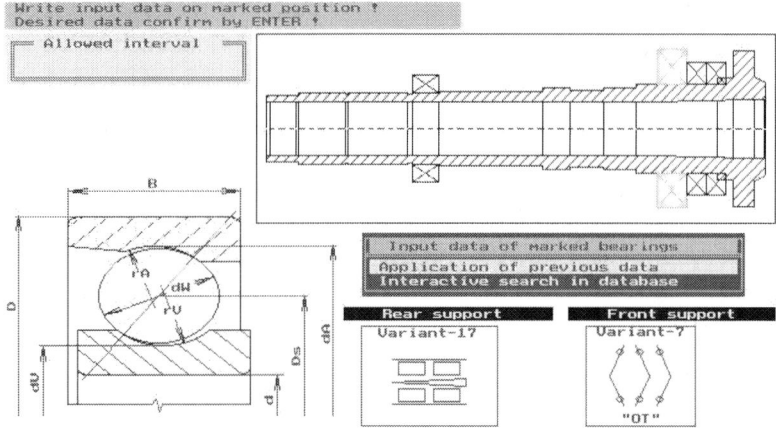

Figure 18. Dimensional design data of the spindle housing

OPTIMIZATION OF THE SPINDLE-BEARING SYSTEM IN RELATION TO TEMPERATURE

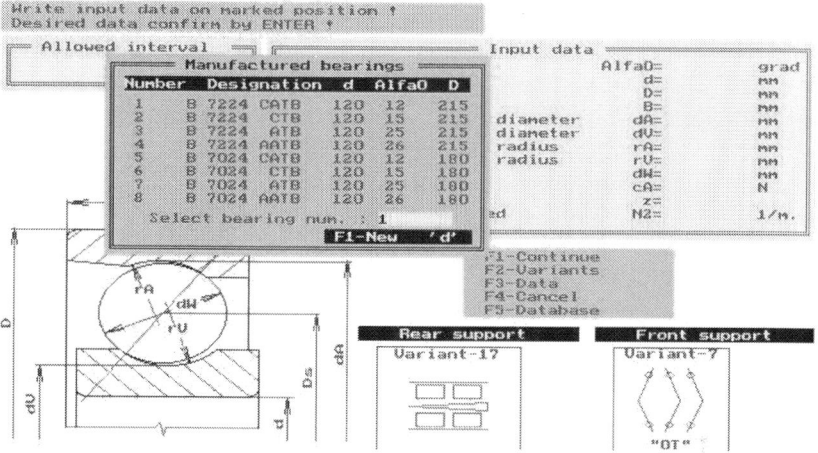

Figure 19. Changing data of the bearings mounting

1. The identification and selection of the standardized spindle nose for turning, milling, grinding and boring;
2. The choice of the design parameters and spindle suitability for different working conditions (working accuracy, preloading, flange type, lubrication system, cooling), Figure 20;

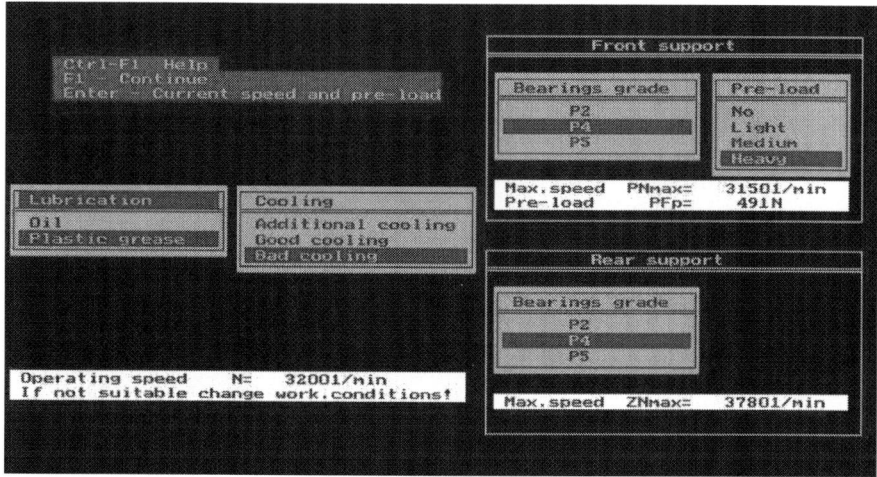

Figure 20. Entering preliminary data for the bearings conditions

1. The calculation and optimization of the cutting parameters for the required material to be machined (cutting force, torque, feed, power), Figure 21;
2. Calculation and optimization of the design and fitment with regard to the applied conditions (revolving speed, radial stiffness, axial stiffness, rating life) for the bearing units and the fitting as a whole, for all of the identified bearing types.

Graphical output of partial deflections caused by bearing nodes distance are shown in Figure 21.

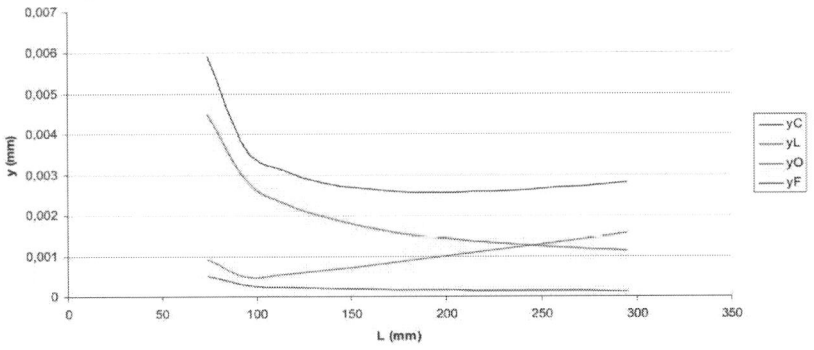

Figure 21. Graphical output of the dependence of partial deflection on bearing node distance

The results include:

- applied entries
- chosen bearing (unit) fit types
- rotational speed limit
- radial unit stiffness
- axial unit stiffness
- resultant stiffness for the chosen spindle fit system
- load parameter and durability
- graphical illustration of the chosen design (Table 1.).

The "Spindle Headstock" software application product is a basic programme which can be modified according to the user's wishes.

The programme has been written in the source code programming language, T-PASCAL v.7, with special additional modules for graphics. A number of interactive modules prepare user-specified data for use in AutoCAD utilising the DXF format. The programme can be used on any IBM/PC compatible computer using a HERCULES, EGA or VGA graphics adapter.

The applied software technology has been used in the industry to improve the working accuracy of the machine tools made by TOS Trenčín-Slovakia, for SN and SPSI type lathes, (2), and to design the boring headstocks for the modular single-purpose machine tools made by TOS Kuřim-Czechoslovakia, (5) TOS Lipník, SKF, GMN and INA Skalica. The programme is very effective and reliable and comparison of the results between experiments and calculations show good correlation, never exceeding 10 %.

EXPERIMENTAL RESEARCH

Theoretical results and hypotheses must be verified by experimental tests.

Research of Bearing Nodes Characteristics
In some cases it is very difficult, or even impossible, to gain experimental results from actual machine tools. This led us to develop an experimental device for research into spindle bearing node arrangement characteristics. Our department has developed such a device that can measure:

1. changes in the bearing contact angle at its mounting point, changes in loading, and changes in revolving frequency,
2. deformation from axial and radial loading for various preload arrangements, contact angles and bearing node revolving speed settings
3. increases in the temperature in the bearing nodes at various settings
4. dimension of cutting forces in bearing nodes

The variation in the stiffness of the bearing arrangement B7216 is shown in Figure 22, [19]. We can use the experimental measuring head for measuring the deflection and temperature of a varying number of bearings and bearing nodes (from 2 to 5), their preload value, dimensions, and the contact angle of the bearings with different radial and axial forces used, [19], [20].

We use this device to measure the deformation characteristics of the bearing node with different combinations of bearing arrangement, pre-stressed values, contact angles, loads and revolution frequencies. We use this experimental measuring head for verifying the theoretical calculation and real performances of the bearing node (stiffness, precision running and temperature).

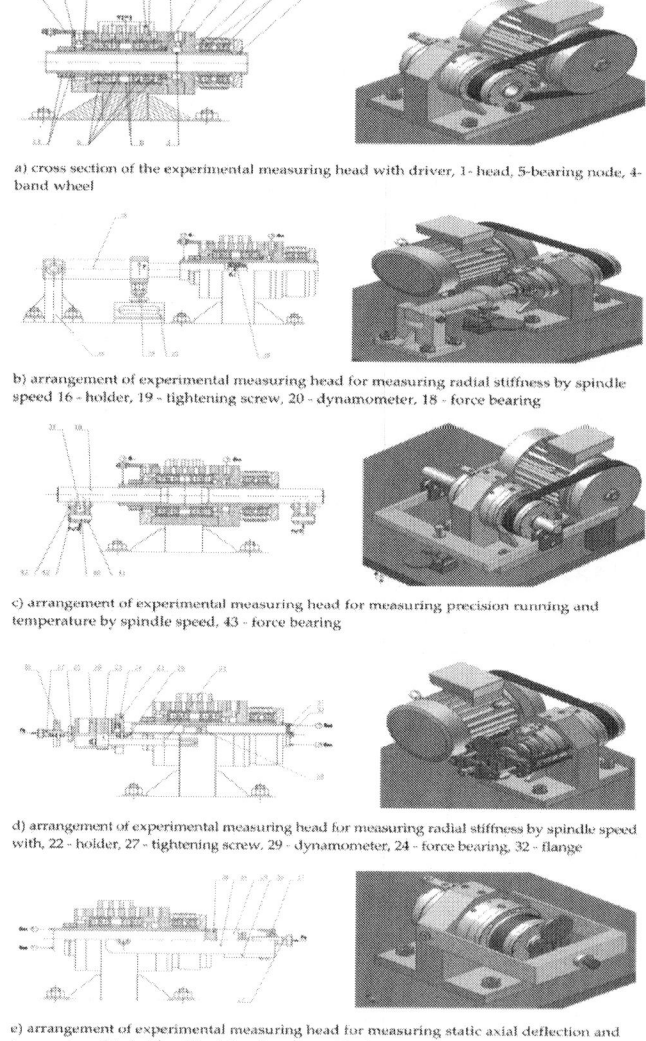

a) cross section of the experimental measuring head with driver, 1- head, 5-bearing node, 4- band wheel

b) arrangement of experimental measuring head for measuring radial stiffness by spindle speed 16 - holder, 19 - tightening screw, 20 - dynamometer, 18 - force bearing

c) arrangement of experimental measuring head for measuring precision running and temperature by spindle speed, 43 - force bearing

d) arrangement of experimental measuring head for measuring radial stiffness by spindle speed with, 22 - holder, 27 - tightening screw, 29 - dynamometer, 24 - force bearing, 32 - flange

e) arrangement of experimental measuring head for measuring static axial deflection and temperature, 34 - holder, 27 - tightening screw, 29 - dynamometer

Figure 22. Different variants of the experimental measuring head.

In Figure 23 we have compared the experimental stiffness measurement, with the accurate theoretical and simplified average calculated radial stiffness of the B7216 AATBP4OUL bearing arrangement. The stiffness variation was examined with a 25% contact angle with nominal value of bearing arrangements: z_1, $z_2 = 14$, d_{w1}, $d_{w2} = 19,05$ mm, α_1, $\alpha_2 = 12°$, $F_p = 340$ N.

When static, the experimental values of radial stiffness are higher than the theoretical values. The dependence of stiffness on loading exhibits a decreasing pattern. The decrease is nearly linear, until a certain critical force "F_{kr}" is reached, at which point the roller with the lightest load becomes unloaded. The deformation characteristic of the nodal point is influenced by the type of flange. The degree and gradation of the stiffness change under the given operational conditions depend on their construction.

In this field the results of the precise and the approximate mathematical model are practically the same. Consequently it follows that in a preliminary mounting design, a simplified mathematical model for calculating the stiffness of the nodal points can be used, as suggested in this article. The convergence of the measured and calculated values provides good evidence for a wider application of the programme in practice.

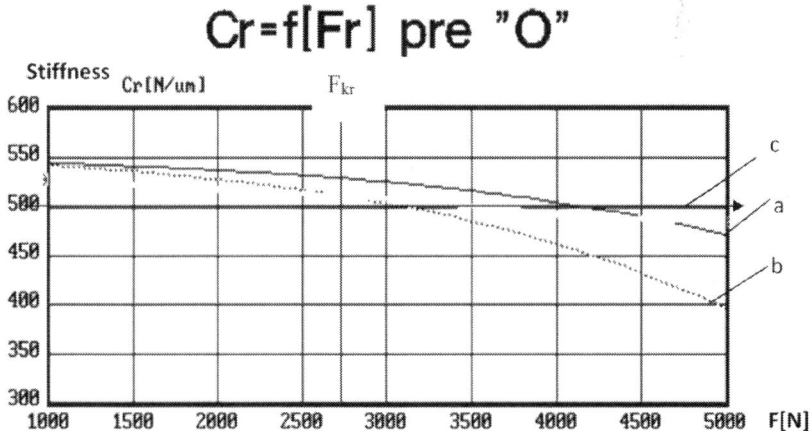

Figure 23. Radial stiffness of the bearing arrangement B7216 AATB P4 O UL, a - experimental, b - accurate theoretical c - simplified average

New. Design of headstock

In the new design of a headstock which connects to a CNC system, the maximum width of cut is limited by the point at which self-exciting vibration starts.

From a constructional point of view, the headstock design can be classified as follows:

- classical headstock
- headstock with an integrated drive unit

The classical headstock is a mechanical unit, where a spindle is driven by a motor through a gearbox without any control system.

The disadvantages of the classical construction are as follows:

- problems with the gears at higher revolving frequencies,
- actual cutting speeds are not continual because of the discontinuous nature of the gearboxes,
- large dimensions of complete units

New design "Duplo–Headstock"

The "Duplo–Headstock" has been designed in order to achieve technological parameters comparable to the performance of standard electro-spindles, but at a lower production costs and with higher controllability. This particular headstock is assembled from readily available elements (bearings, single drives,). The demands on the other peripheral devices are reduced, as are the costs.

The "Duplo-headstock" can be described as a spindle with double supports, driven by two separate motors which can operate independently or together. Figure 24- 28 show a „Duplo–headstock" design [20].

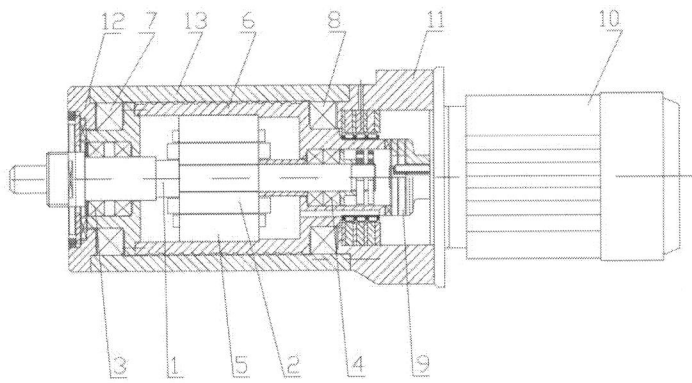

Figure 24. High-speed headstock "Duplo"

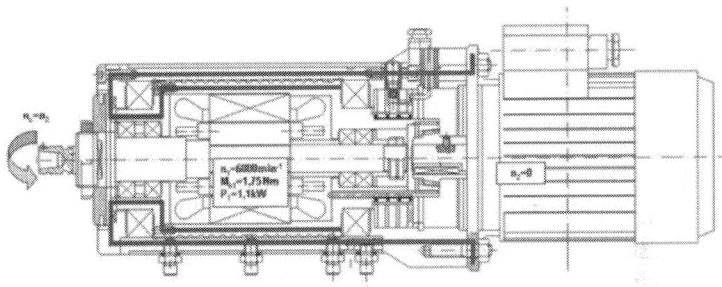

Figure 25. The stator engaged on spindle, Speed: n_{1max}=6000 (min-1); n_2=0, n_{c_sp}=n_1 Torque moment: Mk_{c_sp}=Mk_1=1,75 (Nm) by n_{1max}, Power: P_{c_sp}=P_1=1,1 (kW)

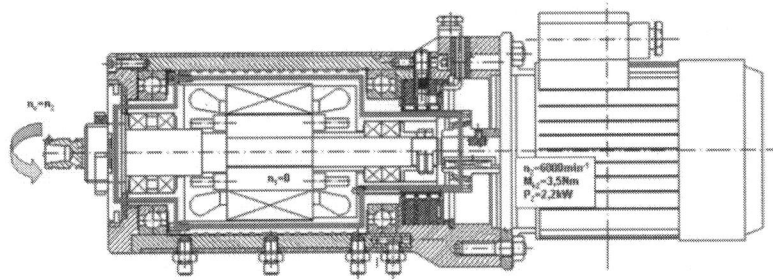

Figure 26. The stator engaged on body, Speed: n_{1max}=0; n_2=6000 (min-1), n_{c_sp}=n_2, Torque moment: Mk_{c_sp}=Mk_2=3,5 (Nm) by n_{2max}, Power: P_{c_sp}=P_2=3,5 (kW)

Machine Component Design

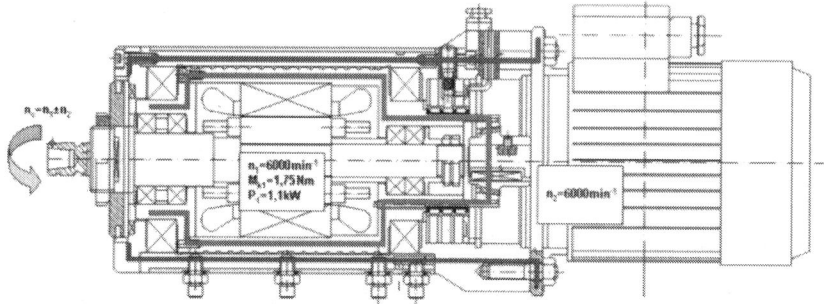

Figure 27. Disengaged, Speed: n_{1max}=6000 (min-1); n_2=6000 (min-1), n_{c_sp}= n_1 + n_2 by one direction of rotation, n_{c_sp}= n_1 - n_2 by opposite direction of rotation, Torque moment: Mk_{c_sp}=Mk_1=1,75 (Nm) by n_{1max}, Power: P_{c_sp}=P_1=1,1 (kW)

Figure 28. Stand of "Duplo" Headstock

The spindle (1), with built-in armature (2), is supported by bearings (3), (4). The stator of the internal motor (5) is supported on bearings (7), (8). The clutch (9) connects a hollow shaft with an external electro-motor (10). The stator feeding rings (11) are located in the rear part of the shaft. The clutch (12) enabling switching between working modes is located in the front part of the shaft. The advantage of this innovative design, which is already in use, is that the headstock can work in three different modes:

- stator is engaged on the spindle
- stator is engaged on the body
- no engagement

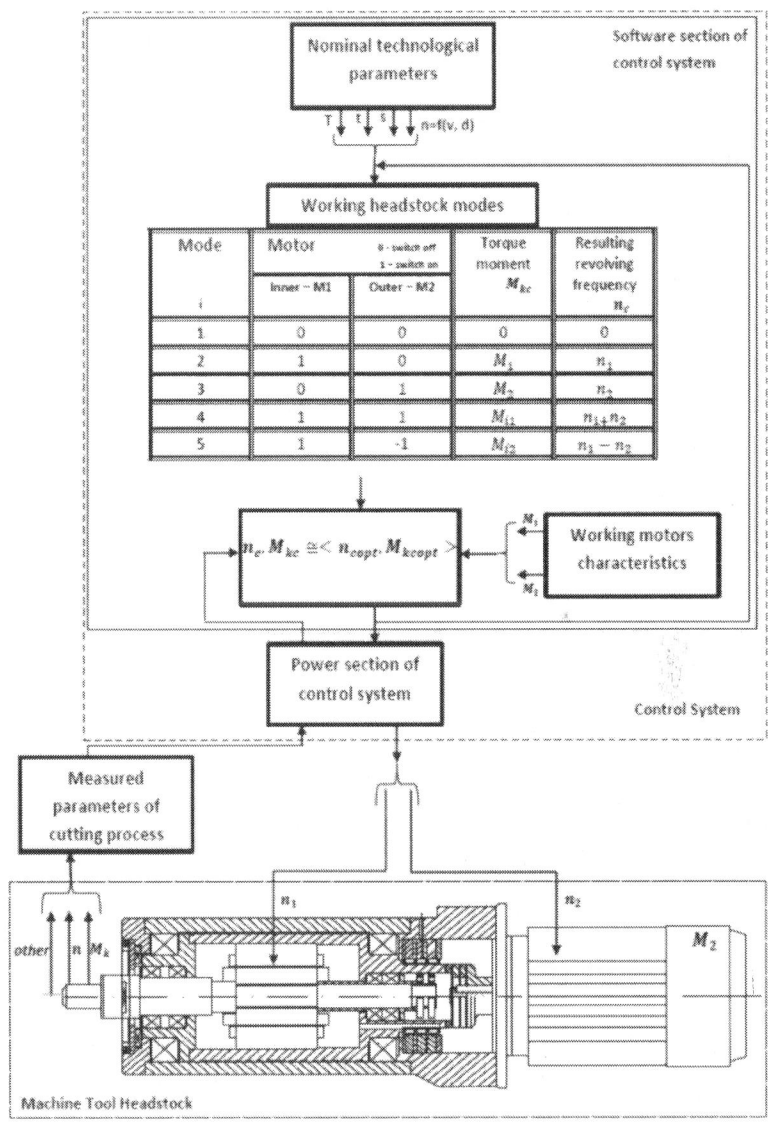

Figure 29. Scheme of Machine Tool Headstock Control

Connecting such a headstock with a suitable control system can provide optimal cutting conditions for various technological operations. The intelligent control system, Figure 29, can operate in any one of the working modes and ensure nominal or optimal technological parameters best suited to the machining process, [21]. Figure 30 shows the design for the construction of the "Duplo" Headstock.

In the third mode, the clutch (12) is switched off. The spindle (1) is driven by both motors, (Figure. 27), providing the maximum speed, which is required, for example, in grinding.

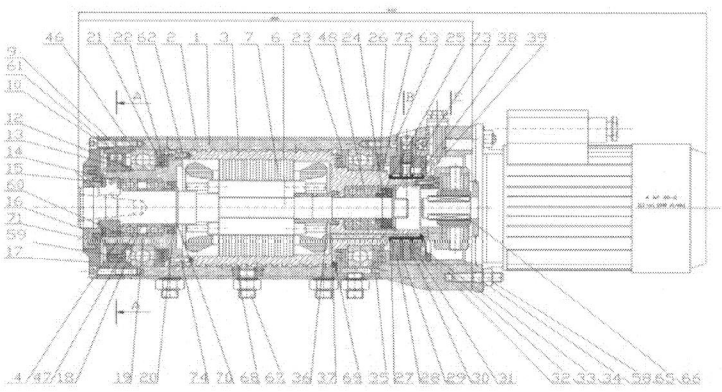

Figure 30. Design for the construction of "Duplo" Headstock

APPLICATION RESEARCH

The New Headstock Construction For Turning Machine Tools
TRENS a. s. Trenčín, is a Slovak manufacturer of machine tools (mainly lathes) and offers a new generation of lathes implementing various technological advances in design, production, and control systems, [22]. The Department of Production Engineering has been asked to design an accurate running spindle for the SBL 500 CNC lathe (Figures 31 -33), [23]. All construction data and results of measurements were obtained from the producer. Table 2.showsthe calculated (Spindle Headstock Version 2.8) values [23] of the optimized design. Figure 34 shows the comparison between the original and optimized designs.

APPLICATION RESEARCH 161

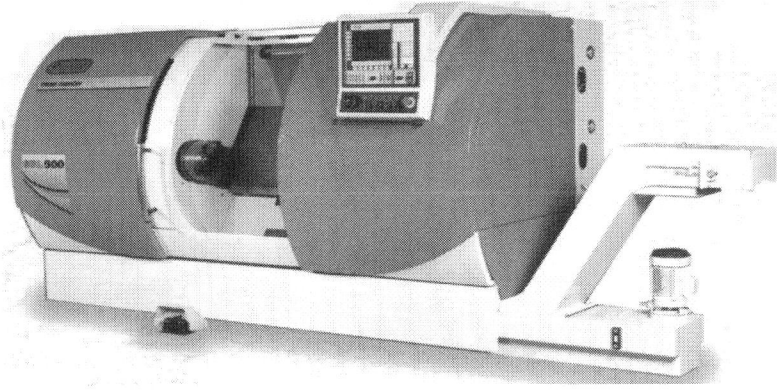

Figure 31. CNC Lathe SBL 500

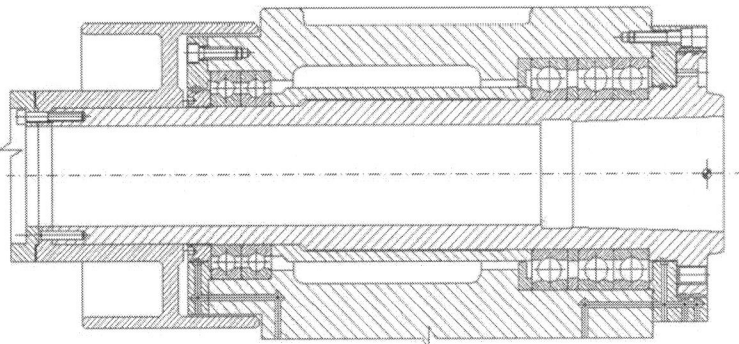

Figure 32. Original design of SBL 500

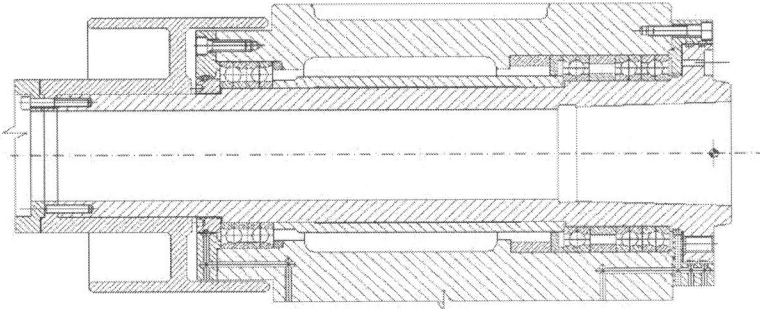

Figure 33. Optimized design of SBL 500

Table 2. Calculated values of optimized design

		Unit	Value	Notice [%]
Total axial stiffness	C_a	[N/μm]	372	
Total radial stiffness	C_r	[N/μm]	351	
Total spindle displacement y_r		[μm]	18,45	
displacement forces resulting from		[μm]		
- the bending moments	y_{Mo}	[μm]	9,79	53,0
- the bearing compliance	y_L	[μm]	6,16	33,5
- the skidding	y_t		2,49	13,5
Limited frequency of rotation n_c		[min^{-1}]	2695	unfit
Life-time	T_h	[hour]	5175	unfit
Distance between supports	L	[mm]	327	

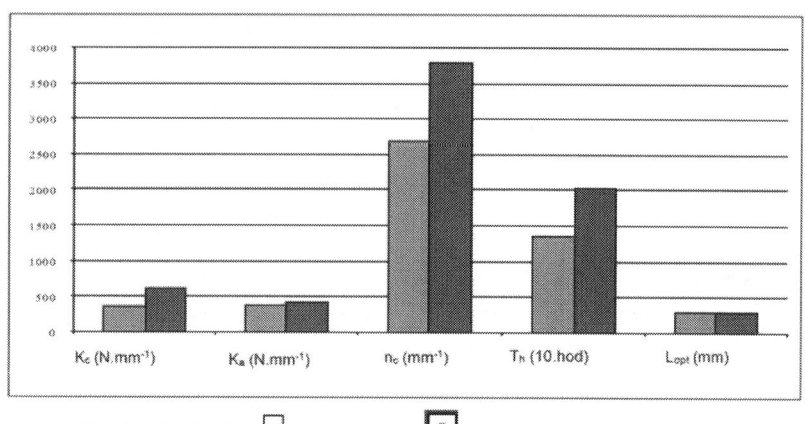

Headstock of SBL 500, ☐ Original design, ■ Optimized design

Figure 34.

Dynamic Analysis

The most valuable advantage of this system is the possibility of calculating dynamic stiffness at different revolving frequencies of the spindle. The given mathematical model was verified on a number of spindles with programs which enabled the calculation of natural frequencies (COSMOS). The results were in good compliance [24].

The verified spindle, which complied with research findings, was reduced to a three discrete parts. The dynamic mathematical model described above was used to calculate the natural frequencies and dynamic deflections. Table 3 compares calculated and experimental values.

Table 3. Experimental and calculated values of frequencies

Frequency	Calculated	Experimental	Difference
f1 (Hz)	1 201	940	+27,8 %
f2(Hz)	1 727	1610	+7,3 %
f3(Hz)	10 605	–	–

The results can be considered as correct, in spite of the relatively large difference in values (28 %) in the first frequency. This is as a resultof the fact that the dimensions of the additional rotating parts are not included. If these parts were included in the calculation, the values of the calculated natural frequencies would be smaller.

An example of the graphic output of calculated values is shown in Figure 35, [23]. The chart shows the dynamic deflection of the spindle reduced to three masses. The first two resonant frequencies of the optimized spindle are marked on the chart.

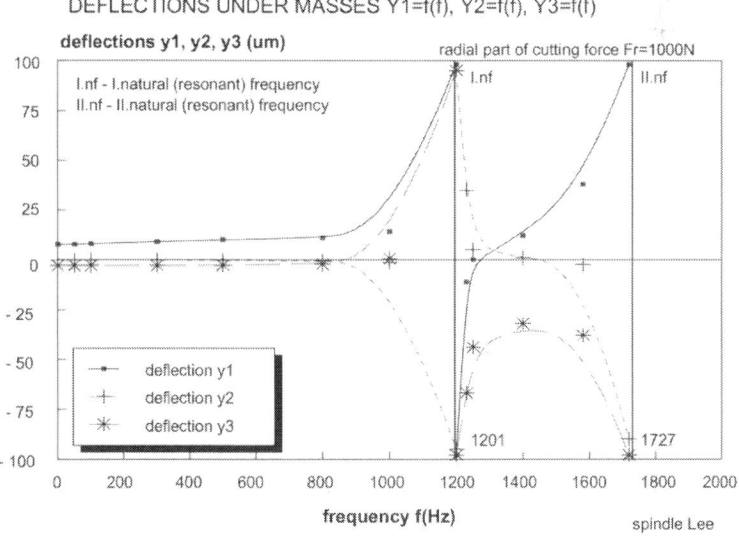

Figure 35. Dynamic deflections of the spindle according to research findings [7]

CONCLUSION

One of the main requirements in designing new spindle housing systems is the ability of the design to be quickly applied to real world practice. The methodologies of calculation that were created must be verified, and models must be adapted into a suitable user friendly, computerized format. The models must illustrate the real characteristics of a spindle housing system.

In this design process, only one variable or parameter was changed and the optimal configuration was identified. The results calculated for a static analysis of the SBL Headstock are presented in Table 2 andFigure 33. The dynamic analysis results are presented in Table 3 and Figure 35. The calculated results were verified with experimental measurements. The difference between measured and calculated values is relatively small.

There is no doubt that the re-design has been a success story, and has proven to be highly effective inthe identification of optimal SBS design. More detailed information can be read in [22], [23] and its application can be seen in the machine tools made by TRENS Inc., The SBL Lathe was presented in the Mechanical Engineering Exhibition in Nitra in 2010 and in the EMO Exhibition in Düseldorf in 2011.

NOMENCLATURE

N - high-speed ability

δ – elastic compression

F – external load

P - roller loading

E - modulus of elasticity of the material

J – quadratic moments of inertia

i – number of bearings

α – contact angle

D, d – diameter

n – high spindle revolutions

l – distance of curvature centre

K - stiffness

γ – pitch angle

O – centre

Indexes

a – axial direction

r radial direction

z – built-in state

l – referring to its bearing

0 – roller loaded to the maximum

j – circumference roller

A – external ring

I – internal ring

m – medium value

w - roller bearing

REFERENCES

1. M.. Weck, N. Hennes, M. Krell, (1999, Spindlel and Toolsystems with High Damping. In: Cirp Annals-manufacturing Technology- CIRP ANN-MANUF. TECHNOL, 4812973021999
2. J. a. Marek, kol, C. N. C. Konstrukce, strojú. obráběcích, s.r.o. Publisching, Praha, 2010978-8-02547-980-3s p.
3. D. Lee, H. Sin, N. Sun, 1985Manufacturing of a Graphite Epoxy Composite Spindle for a Machine ToolCIRP, 34, 1365369
4. Ľ. Šooš, 2008Spindle headstock- the heart of machine tool.In: Machine Design: On the occasion of 48th anniversary of the Faculty of Technical Sciences: 1960-2008, Novi Sad: University of Novi Sad. 978-8-67892-105-6335340
5. Ľ. Šooš, 2008Quality of design engineering: Case of machine tools headstock.In: Quality Festival 2008 : 2nd International quality conference.- Kragujevac, May 1315Kragujevac: University in Kragujevac, 978-8-68666-325-2

6. Ľ. Šooš, 2008Contribution to the research of static and dynamic properties of CNC turning machine In: Strojníckyčasopis = Journal of Mechanical engineering. 0039-2472Roč. 59, č. 5-6 (2008), 231239
7. L. Javorčík, Ľ. Šooš, J. Zon, 1991Applied software technology for designing a bearing housing fitted with rolling bearing arrangemant. in.:"ICED 91". Zurich, August, 1991. 12281233
8. Ľ. Šooš, L. Javorčík, P. Šarkan, 1995An inteligent drive unit for Machine Tools. In.: The first world congress on Intelligent manufacturing porceedings, university of Puerto Rico, 13-17. February 1995, 344352
9. P. Demeč, 2001Presnosťobrábacíchstrojov a jejmatematickémodelovanie.- 1. vyd.- Košice: Technickáuniverzita v Košiciach.- 146 p.- 807099620in Slovak).
10. Ľ. Šooš, 2008Criteria for selection of bearings arrangements. In: 32. SavetovanjeproizvodnogmašinstvaSrbijesamedunarodnimučešcem = 32nd Conference on production engineering of Serbia with foreign participants: Zbornikradova = Proceedings.- Novi Sad, 18.-20. 9. 2008.- Serbia: Fakultettehničkihnauka.- 978-8-67892-131-5395399p.
11. Ľ. Šooš, 2010New methodology calculations of radial stiffness nodal points spindle machine tool. In: International symposium on Advanced Engineering & Applied Management- 40th Anniversary in Higher Education: Romania /Hunedoara/ 45November, 2010.- Hunedoara: Faculty of Engineering Hunedoara.- 978-9-73009-340-7III-99- III-104.
12. Ľ. Šooš, 2011Approximate methodology calculations of stiffness nodal points.In: World Academy of Science, Engineering and Technology.- 0201-0376X.- Year 7, 8013901395
13. T. A. Harris, (196, 1966Rolling Bearing AnalysisNew York- London-Sydney, 1966, 481 ps.
14. V. B. Balmont, S. P. Russkich, 1978Rasčetradialnojžestkostiradialno-upornogopodšipnika. Trudy instituta. M., SpecinformcentrVNIPPa, 69, 1978,č.1, s..93107
15. Kovalev,M.P.-Narodeckiij,M.Z.:(1980Rasčetvysokotočnychšarikopodšipnikov. 2 vyd. Moskva, Mašinostroenie 1980. 279s p.
16. Ľ. Šooš, 2008Radial stiffness of nodal points of a spindle. In: MATAR Praha 2008. Part 2: Testing, technology: Proceedings of international congresss.- Prague 16th-17th September, Brno 18th September 2008.- Praha: Českévysokéučenítechnické v Praze- 978-8-09040-770-14347
17. Ľ. Šooš, B. Á. B. I. C. Sábics, J. , (198, 1989Axiálnatuhosťvysokootáčko výchvretienobrábacíchstrojov. In.:Strojírenství, 39, 1989, č.2, pp s. 8691

18. Ľ. Šooš, P. Šarkan, 2004Design of spindle-bearing arrangement of angular ball bearings. In.: MMA 94:Fleksibilnetechnologije: 11th International conference on Flexible Technologies. Novi Sad, 8- 9.6.2004.- Novi Sad: Institutzaproizvodnomašinstvo-271275
19. Šooš, Ľ.-Valčuha, Š.- Bábics, J.: PV 08651-88 Zariadenienaskúšaniev alivýchložísk [Patent].
20. Ľ. Šooš, Generátorskladanýchrotačnýchpohybov, 2009ísloúžitkovéhovzoru: SK 5363.- Dátumnadobudnutia: 22. 12. 2009, [Patent].
21. Ľ. Šooš, pohon. Duplo, In. realitaalebovízia?, Mechanica. Acta, Slovaca, I. S. S. N. 13352393 , 1. Roč, č. . , A. . konf, Celoštátnakonferencia. s. (heslo, . medzinárodnouúčasťou, R. O. B. T. E. P. 200, 2006Jasná- NízkeTatry, 31.5.- 2.6.2006 (2006).- Košice : Technickáuniverzita v Košiciach, spp. 515-518
22. Ľ. Šooš, Contribution to the research of static and dynamic properties of CNC turning machine. In: Strojníckyčasopis = Journal of Mechanical engineering.- 0039-2472Roč. 59, č. 5-6 (2008231239
23. Ľ. Šooš, system. S. B. L. 5. Spindle-housing, C. N. C. In, Niezawodnošč. =. Eksploatacjai, Maintenance, reliability, I. S. S. N. 1507-2711, Č. . , 20085356
24. Ľ. Šooš, New methodology calculations of radial stiffness nodal points spindle machine tool. In: International symposium on Advanced Engineering & Applied Management- 40th Anniversary in Higher Education: Romania /Hunedoara/ 45November, 2010.- Hunedoara: Faculty of Engineering Hunedoara, 2010978-9-73009-340-7III-99- III-104.

CITATION

Ľubomír Šooš (2012). Radial Ball Bearings with Angular Contact in Machine Tools, Performance Evaluation of Bearings, Dr. Rakesh Sehgal (Ed.), ISBN: 978-953-51-0786-6, InTech, DOI: 10.5772/51004.

CHAPTER 6

Cryogenic Tribology in High–Speed Bearings and Shaft Seals of Rocket Turbo Pumps

Masataka Nosaka[1] and Takahisa Kato[1]

[1] Department of Mechanical Engineering, University of Tokyo, Tokyo, Japan

INTRODUCTION

In recent years, as a rule, improvement of the reliability of liquid propellant rockets becomes an international technical problem for built-up of safe space transport systems. The high performance, liquid propellant rocket engines require high-pressured turbopumps to deliver extremely low temperature propellants of liquid oxygen (LO_2, boiling point 90 K) and liquid hydrogen (LH_2, boiling point 20 K) to a combustion chamber in engine [1]. In LO_2/LH_2 turbopumps, cryogenic high-speed bearings and rotating-shaft seals are very important parts to sustain high reliability of the high-rotating-shaft systems. The turbopump bearings are directly equipped in cryogenic propellants in pump side [2]. The shaft seal systems are also set up between the cryogenic pumps and the hot turbines to restrain the leakage of cryogenic propellants and hot turbine gas [3].

These bearing and shaft seals have to operate under poor lubricating conditions due to extremely small viscosity at cryogenic temperatures. Furthermore, the turbopump bearings and shaft seals have to overcome a severe high-speed operation that has several critical speeds demonstrating self-induced severe vibration of the rotating shaft. In order to develop turbopump bearings and shaft seals, many inexperienced technical and tribological problems must be solved for extremely low temperature and high speed of operational conditions. Such cryogenic tribological technology has been playing a key role in cryogenic turbo pumps to achieve high reliability.

This chapter presents a topical review of cryogenic tribological studies (for about 30 years in Japan) on the research and development of the cryogenic high-speed bearings and shaft seals of rocket turbopumps [4, 5]. The high-speed bearings and shaft seals were continually studied for the LE-5 engine used in the Japanese H-I rocket (developed in 1986) and the LE-7 engine used in the H-II rocket (developed in 1994). The bearings and shaft seals used in LO_2/LH_2 turbopumps of the LE-5 and LE-7 had a rotational speed level of 50,000 rpm and had been studied and developed from the mid-1970 to the mid-1990. Specially, the all-steel bearings (made of AISI 440C) of the LH_2 turbopump of the LE-7 demonstrated high performance with high reliability at high-speed level at 2 million DN (40 mm x 50,000 rpm). The shaft seal systems in the LE-5/LE-7 turbopumps that used a mechanical seal, a floating ring seal (annular seal) and a segmented seal are also reviewed.

Furthermore, for future space transport systems to reduce launch cost and to increase efficiency, advanced rocket engines which are characterized by high durability (long life) and high performance (light weight) are required in recent years. Advanced bearing and shaft seal that have high durability, i.e., a long life of 7.5 hours for the turbopump bearings used in reusable space shuttle main engine (the SSME). Its required life is 15 times longer than that (30 minutes) of the turbopump bearings used in the LE-7. At the first time, the SSME turbopump bearings experienced a serious wear problem in LO_2 due to poor self-lubrication of the retainer [6]. In order to extend bearing life, the hybrid ceramic bearing with Si_3N_4 balls was used to reduce serious wear in the conventional all-steel bearing. A new type of the retainer having PTFE/bronze-powder insert was also developed to obtain sufficient self-lubrication of the hybrid ceramic bearing. Consequently, the improvement of the SSME turbopump bearings needed a long time of about 20 years [7].

Today, ultra-high speed level above 100,000 rpm is required to make a small and light turbopump for advanced second-stage engine. These advanced research and development are actively underway. In Japan, a new type of hybrid ceramic bearing having Si_3N_4 balls with a single guided retainer demonstrated excellent performance at an ultra-high speed of 120,000 rpm (3 million DN) in LH_2 and recorded the world's top speed (in 2001) [8]. The result of this bearing was applied to the LH_2turbopump (rotational speed, 90,000 rpm) of the RL60 demonstrator engine (in 2003). The RL60 demonstrator engine was developed in the USA with international collaboration (USA, Japan, Russia and Sweden) and the LH$_2$ turbopump was developed by a Japanese company [9]. In Europe, for the VINCI engine under development, high-DN hybrid ceramic bearing was tested in LH$_2$ at a speed of 70,000 rpm (2.8 million DN) and continuous studies on a high-DN bearing was conducted at DN up to 3.3 million (120,000 rpm) in LH$_2$ (in 2005) [10]. Furthermore, in Russia, for the developed RD0146 engine, its rotational speed of the main LH$_2$ turbopump was 123,000 rpm (3.08 million DN), but detail of its bearing was unknown (in 2003) [11].

This chapter also reviews advanced bearings and shaft seals which were studied from the mid-1990 to the mid-2000 after the development of turbopump bearings and shaft seals of the LE-7 [4,5]. It is typical that a long-life bearing with single-guided retainer demonstrated a long operation for 12 hours under 50,000 rpm. A hybrid ceramic bearing having single-guided retainer and Si$_3$N$_4$ balls was able to demonstrate ultra-high-speed performance at speeds up to 120,000 rpm and show excellent performance under 3 million DN. An annular seal made of an Ag plated steel ring also presented two-phase seal performance at speeds up to 120,000 rpm.

These historical reviews are intended to help the technical succession to next young generation who challenges research and development of the future space transportation system. These reviews are based on previous studies carried out by Japan Aerospace Exploration Agency (JAXA) at Kakuda Space Center. All materials used in this chapter have been published by papers.

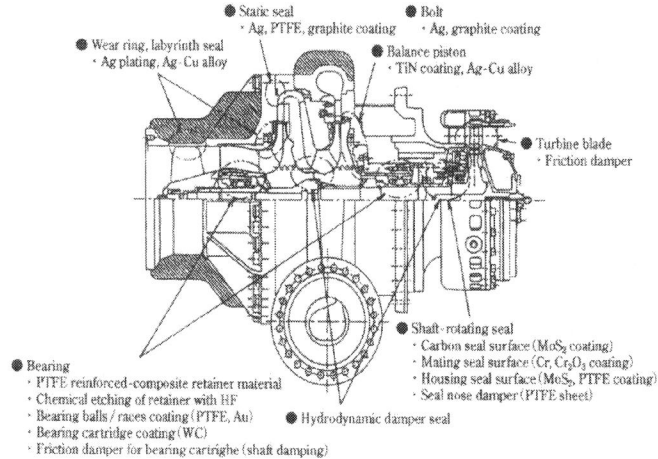

Figure 1. Typical tribo-components and solid lubricants used in turbopumps

BEARINGS AND SHAFT SEALS OF TURBOPUMPS

Turbopumps And Tribo–Components

The LO_2/LH_2 turbopumps as well as the tribo-components, such as high-speed bearings and rotating shaft-seals, were studied and developed to use in the LE-5 and LE-7. In reference to the structure of the LH_2 turbopump of the LE-7, the tribo-components and solid lubricants used in the LE-5 and LE-7 turbopumps are typically indicated in Fig. 1 [4]. In addition, main design parameters of the turbopumps and DN values of bearings for the LE-5 and LE-7 are listed in Table 1 [5]. Here, the DN value that represents high-speed level of bearing is defined as the product of the inner-race bore diameter D (in mm) and the pump rotational speed N (in rpm). The rotor speed is typically restricted by the DN limits of the bearing.

Table 1. Design parameters of turbopumps and DN values of bearings for LE-5 and LE-7

Engine (thrust)	LE-5 (10 tons)		LE-7 (86 tons)	
Rocket	Second stage of H-1		First stage of H-2	
Engine cycle	Gas-generator cycle		Staged-combustion cycle	
Turbopump	LO_2	LH_2	LO_2	LH_2
Pump pressure [MPa]	5.2	5.5	17.4 (25.8)*	27
Pump flow rate [kg/s]	20	3.6	212 (46)*	36
Shaft rotational speed [rpm]	16,500	50,000	18,000	42,000
Bearing DN [mm x rpm]	49.5×10^4	125×10^4	81×10^4	168×10^4
Turbine pressure [MPa]	0.48	2.4	19.1	20.6
Turbine temperature [K]	690	840	810	830
Turbine gas flow rate [kg/s]	0.39	0.42	14.9	33.6
Shaft power [kW]	130	490	4,700	19,700
Weight [kg]	23	25	160	200

[i] - For pre-burner in bracket; ( )*

a. *LE-5 turbopumps*

For the upper stage of the H-I rocket, the LE-5 had a gas-generator cycle with 10-ton thrust and its chamber pressure of 3.4 MPa was relatively low. Its engine cycle is not able to achieve a high engine performance due to an open cycle. For the LH_2 turbopump of the LE-5, the pump discharge pressure was relatively low at 5.5 MPa and the discharge flow rate was 51 liters/s. The turbine pressure was 2.4 MPa. The paired bearings of 25-mm bore operated at a speed of 50,000 rpm (1.25 million DN) and sustained the shaft power of 490 kW [12].

For the LO_2 turbopumps, the discharge pressure was 5.2 MPa and the discharge flow rate was 18 liters/s. The turbine pressure was 0.48 MPa. The paired bearings of 30-mm bore operated at a speed of 16,500 rpm and sustained the shaft power of 130 kW. Basic design and technology of the cryogenic tribo-components used in the small turbopumps was experimentally established under the development of the LE-5.

a. *LE-7 turbopumps*

For next technical challenge in the first stage engine of the H-II rocket, the LE-7 had a staged-combustion cycle (similar to that of the SSME) with 100-ton thrust and a high chamber pressure of 13 MPa. Its engine cycle can obtain high performance due to a closed engine cycle. For the high-pressure, large LH_2 turbopump of the LE-7, the pump discharge pressure was increased to 27 MPa, and the discharge LH_2 flow rate was 510 liters/s. The turbine pressure was relatively high at 20.6 MPa. The paired bearings of 35-mm bore were at the inducer side, and the paired bearings of 40-mm bore were at the turbine side. These bearings operated at a speed of 42,000 rpm (1.68 million DN) and sustained the shaft power of 19,700 kW [13,14].

For the LO_2 turbopumps, the discharge pressure was 18 MPa for the main pump and 26 MPa for the preburner pump, respectively. The total discharge LO_2 flow rate was 240 liters/s. The turbine pressure was 19.1 MPa. The paired bearings of 32-mm bore were located at the inducer side and the paired bearings of 45-mm bore were at the turbine side. These bearings operated at a speed of 18,000 rpm and sustained the shaft power of 4,700 kW [14,15].

a. *Tribo-components in turbopumps*

As shown in Fig. 1, it is important to prohibit severe friction and wear in cryogenic environment that various solid lubricants are applied to the frictional parts in static and dynamic tribo-components. Since the turbopums are operated under large power conditions connecting with high fluid and mechanical vibration, it must pay attention that many components in contact are sure to generate relative motion and resulted in severe adhesive conditions. It needs proper lubrication to avoid severe frictional adhesion of assembled parts used in cryogenic environment.

The rotor of turbopump is directly supported by two sets of self-lubricated ball bearings in cryogenic pump fluid. The shaft seal of turbopump is installed between the cryogenic pump and the hot gas turbine. The shaft seal system must seal the cryogenic propellants and the combustion gases (steam with rich hydrogen gas) safely and securely. High-speed components, such as bearings, shaft seals, Labyrinth seals, wear rings and balance pistons, used the proper solid lubricants to protect them from severe friction and wear in the reduction (LH_2) or oxidation (LO_2) environment of the cryogenic propellants. It is noted that these high-speed tribo-components are important life-controlling parts in engines [4].

SELF–LUBRICATING BEARINGS

Self–Lubrication of Retainer

The turbopump bearings are all-steel (AISI 440C) bearings that are self-lubricated by the PTFE transfer film as a lubricant from the reinforced PTFE (polytetra fluoroethylene) retainer. AISI 440C is martensitic stainless steel (with 16-18%Cr) and is one of the most widely used bearing materials in space systems because such high-Cr steel has a high corrosion resistance due to a superficial surface layer of Cr_2O_3. The resin PTFE retainer is reinforced with glass fiber, carbon fiber and laminated glass cloth to reduce wear as well as thermal contraction of the retainer. Although PTFE material has poor mechanical strength at room temperature, it has the best lubricant for use at cryogenic temperature because its mechanical tensile stress drastically increases and reaches to 80 MPa in LO_2 and 130 MPa in LH_2, respectively. In order to reduce wear of the PTFE composite retainer with poor thermal conductivity, sufficient cooling of the retainer is need to eliminate heat generation detrimental to successful bearing operation at high speeds [16].

Since LH_2 and LO_2 are particularly poor as lubricants because of their low viscosity under conditions of reduction or oxidation, hydrodynamic fluid lubrication is less effective. It is noted that the cryogenic pump fluids works to remove severe frictional heat and to prevent the temperature rise in the bearing. At low temperatures, the PTFE transfer film as a lubricant is kept to be hard and to sustain the bearing load, so that softening and rupturing of the transfer film due to a rise in temperature have to be eliminated. Under poor cooling conditions, it appears that the blackened transfer film due to thermal decomposition of PTFE should occur at a high temperature above about 500 K, and the degraded transfer film was not able to sustain the bearing load. Therefore, sufficient cooling by cryogenic fluids, as well as reduction of frictional heat generation, is very important to produce a durable lubricant film transferred from the retainer even in cryogenic fluid [14].

High–Speed and Load Conditions Of Bearing

For the turbopump bearings, angular-contact bearings are usually used in pairs in duplex mounts (back to back). For example, Table 2 shows main design parameters and internal load conditions for the bearings used in the LH_2 turbopumps of the LE-5 and LE-7 [17,18]. In this table, the *SVmax* value (=*Smax* x *Vmax*/2) that represents the maximum product of stress times spinning velocity in the contact ellipse zone at the inner race are shown. Here, *Smax* is the maximum contact stress and *Vmax* is the maximum spinning velocity. The *SVmax* value is an important factor related to lubrication and wear at the inner race with ball spinning [13,19]. High *SVmax* value leads to high frictional heating and to wear of the PTFE transfer film due to spin wear. Under poor cooling condition and large tilted misalignment, the turbopump bearings have an initial contact angel of 15-25 deg. with a large radial clearance to prevent a loss of operating clearance from bearing seizer. As mention later, high-speed bearing has the outer-race ball control that produces high ball spinning at the inner race. In order to reduce the stress level within the spinning contact zone, race curvatures were controlled to be 0.54-0.56 for inner race and 0.52 for the outer race, respectively. The inner race has a counter-bore type to gain sufficient cooling within the bearing.

As the centrifugal force developed on the balls increases at high speeds, the operational contact angle at the inner and the outer races are changed to be different each other. The operational contact angle at the inner race increases rather than the initial contact angle and decreases to near zero at the outer race. This divergence of contact angles tends to increase ball spinning in addition to rolling at the inner race. Its spin velocity due to ball spinning becomes high and results in an occurrence of frictional heat generation. To contrast, rolling contact at the outer race generates differential slip due to curvature of contact ellipse [20].

TABLE 2. Design parameters and internal load conditions for LH_2 turbopump bearings of LE-5 and LE-7

Parameters	LE-5	LE-7
Bearing		
Dimension [mm]	25 x 52 x 15	40 x 70 x 16
Pitch diameter [mm]	38.5	57
Ball diameter [mm]	7.938	9.525
Number of balls	11	13
Initial contact angle [deg.]	20	25
Initial radial clearance [μm]	57	137
Operating condition		
Rotational speed [rpm]	50,000	46,000
Thrust pre-load [N]	784	1,176
Bearing DN [mm x rpm]	125×10^4	184×10^4
Internal load condition		
Normal load at inner / outer races [N]	157 / 343	176 / 637
Maximum contact stress at inner / outer races (S_{max}) [GPa]	1.58 / 1.49	1.54 / 1.63
Maximum SV at inner race (SV_{max}) [N/mm² x m/s]	2.4×10^3	3.1×10^3

Under the outer-race control connected with ball spinning at the inner race, heat generation due to ball spin is significantly higher than that of differentia slip, so that sufficient cooling is necessary at the inner race side. Furthermore, sliding velocity of the rolling balls in contact with the outer guide land and the ball pocket is high and resulted in a generation of frictional heating of the retainer. The bearings were effectively cooled by the pump cryogenic fluids circulating in the turbopumps. For example, Fig. 2 shows sliding frictional conditions of the inner and outer raceways for the 25-mm-bore bearing that is at a speed of 50,000 rpm under a thrust load of 980 N [16]. This bearing was used in the LH_2 turbopump bearing for the LE-5. In this figure, the distribution of the contact stress, the spinning velocity and the *SV* value with spin at the inner race are shown. Pattern of spin wear generated by ball spinning becomes similar to the distribution of the *SV* value. To contrast, for the outer race, the differential slip velocity and the *SV* value with differential slip are light so that wear due to differential slip is small. Furthermore, for the retainer, the sliding velocity is 50 m/s at the ball pocket and 45 m/s at the outer guide land at a speed of 50,000 rpm, respectively.

For system design of the turbopump high-speed rotor, the thrust load applied on the rotor due to unbalanced fluid pressures is balanced automatically by a balance piston mechanism during operation [17]. As a result, the turbopump bearings can operate only with a spring thrust load to remove internal clearance and control radial stiffness. However, the shaft vibration as well as the fluid action around the impeller should add high dynamic radial load to the thrust load on the bearing. For example, the LH_2 turbopump bearings of the LE-7 had to operate at a speed of 42,000 rpm that was beyond the third critical speed of 32,000 rpm and must support the high shaft-power under high shaft-vibration. Therefore, the bearings must have high combined radial and thrust load capacity at all extremes of the tutbopump operating conditions [14].

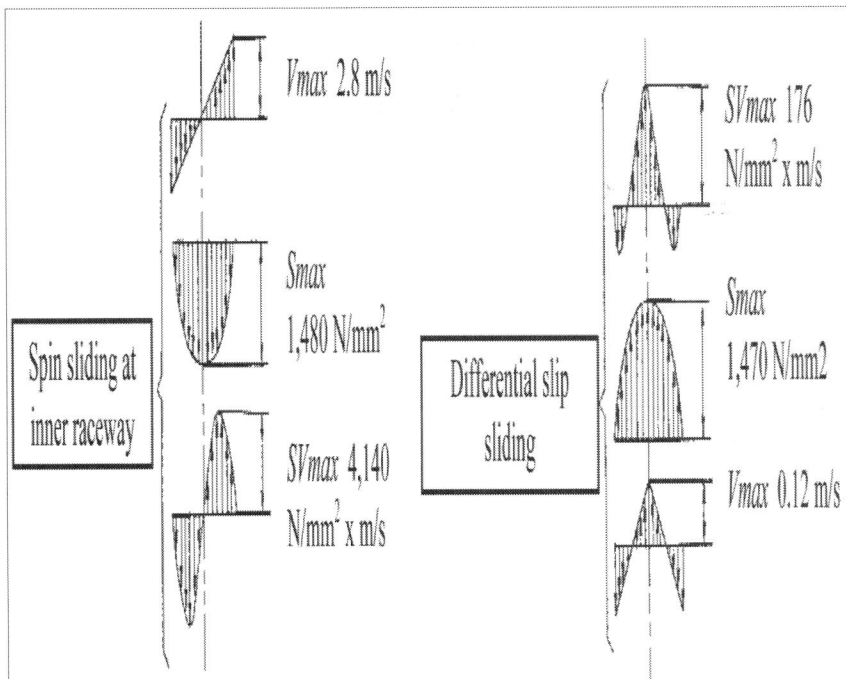

Figure 2. Sliding frictional conditions at inner and outer raceways for LH_2 bearing (25-mm bore, 50,000 rpm, 980 N)

Shaft Seal Systems

The required functions for shaft seal systems vary for different engine cycles. Similar to the SSME, the LE-7 has a two-stage combustion cycle. It requires a high pressure seal since the pressure in the pump and turbine is extremely high. To contrast, the pressure of the pump and turbine in the LE-5 with a gas generation cycle is comparatively low. Design parameters (the seal diameter, rubbing speed and seal pressure) for the seal elements used in the LE-5 and LE-7 turbopumps are listed in Table 3 [21]. The seal elements are the LO_2 seal, LH_2 seal, gas helium (GHe) purge seal and turbine gas seal. These shaft seals prevent or minimize the leakage of LO_2 and LH_2 for pump side and hot turbine gas (steam with rich hydrogen gas) for turbine side. In order to make a short length of the shaft, the shaft seals have to be compactly installed between the cryogenic pump and hot turbine.

Table 3. Design parameters for seal elements used in LE-5 and LE-7 turbopumps

Parameters	Seal diameter [mm]—Rubbing velocity [m/s] (Rotating speed [rpm])— Seal pressure [MPa]— Seal type	
Engine	LE-5	LE-7
LO_2 seal	46.6—40 (16,500)—0.98—(a)	55—58 (18,000)—4.9—(b)
LH_2 seal	43.2—113 (50,000)—1.4—(a)	50—120 (42,000)—7.1—(c)
GHe purge seal	40—35 (16,500)—0.3—(b)	69—173 (42,000)—0.6—(d)
Turbine gas seal	70—61 (16,500)—0.3—(b)	100—105 (18,000)—0.6—(b)
		55—58 (18,000)—16.7—(c)

[i] - (a) Mechanical seal, (b) Segmented seal, (c) Floating ring seal, (d) Lift-off seal

For the LO_2 turbopumps, when the leakage of LO_2 and hot turbine gas are mixed, an explosion will occur. In order to separate the leakage of LO_2 and hot turbine gas in safety, the system is complicated and requires three types of seal elements (the LO_2 seal, GHe purge seal and turbine gas seal). The GHe purge seal installed between the LO_2 seal and turbine gas seal supplies GHe as a barrier gas. To contrast, for the LH_2 turbopumps, the LH_2 leakage can be discharged to the turbine side so that the seal system is relatively simple. However, the rubbing speed of the seal face becomes considerably high and the contacting seal face is opposite severe tribological condition.

For the low-pressure turbopumps of the LE-5, the LO_2 and LH_2 seals used face-contact mechanical seals to gain small leakage. The GHe purge seal and turbine gas seal used contact-type segmented seal. For the high-pressure turbopumps of the LE-7, the LO_2 seal, LH_2 seal and turbine gas seal used non-contact type, floating-ring seal (annular seal) due to high seal pressure. For example, the shaft seal system of the high- pressure LO_2 turbopump of the LE-7 is shown in Fig. 3 [22]. The shaft seal system was set up between the cryogenic pumps and the hot turbine and prevented the mixing of the leakage of LO_2 and hot turbine gas. The LO_2 seal was composed of a floating-ring seal. The turbine gas seal used two floating-ring seals to seal the low temperature GH_2 that made a barrier to the turbine hot gas. So that the turbine gas seal was kept at a lower temperature against the hot turbine section and the reliability of the shaft-seal system was further increased. Between the LO_2 seal and the turbine gas seal, the segmented circumferential seal (GHe purge seal), that had shrouded Rayleigh step hydrodynamic lift-pads to increase opening force, was paired and purged with GHe to prevent mixing of the leakage of LO_2 and GH_2.

The LH_2 seal system of the high pressured LH_2 turbopump was assembled with the floating-ring seal and lift-off seal. The lift-off seal is similar to a face-contact mechanical seal and is in contact with the mating ring (rotating seal-ring) and its leakage is small when the seal pressure is low. As the rotational speed of the turbopump increases and the seal pressure becomes high, the seal faces are automatically disengaged from contacting and changed to be non-contact seal [21].

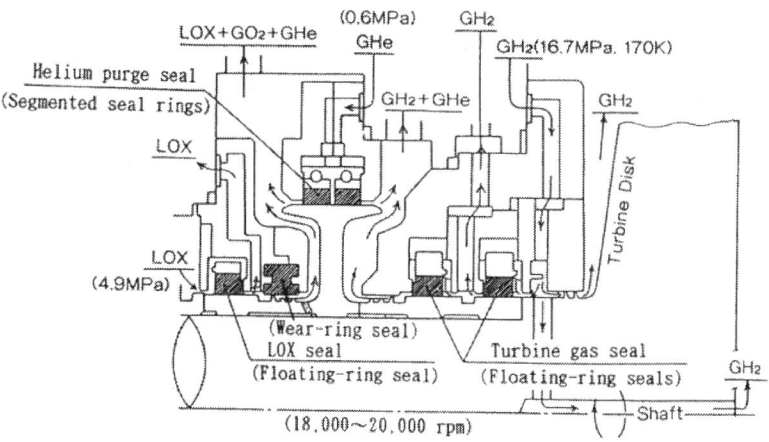

Figure 3. Shaft seal system for high-pressure LO_2 turbopump of LE-7

CRYOGENIC TRIBOLOGICAL PROBLEMS [4, 16, 20, 23]

LH_2 is a particularly poor lubricant due to its extremely low viscosity (approximately equal to that of room-temperature air) and chemical reducing effect to remove native oxide film and to make fresh frictional surface, resulting in a severe lubricating condition at the frictional interfaces. Furthermore, at extremely low temperatures in LH_2, the specific heats and thermal conductivities of tribo-materials drop off rapidly rather than those at the liquid nitrogen (LN_2, boiling point 77 K) temperature. At a high temperature in LN_2, the specific heats and thermal conductivities are less changed and same as those at a room temperature. In addition with vaporization of LH_2, it is easy to produce local hot spots at frictional interfaces, so that frictional condition resulted in severe adhesive (welding) wear in LH_2.

LO_2 has high oxidization power and forms oxide film at frictional surfaces, so that oxide film produces lower friction compared with that in LH_2; however, in boiling of LO_2, oxide wear should increase due to high oxidization power. Active cooling is important to prohibit boiling of LO_2 at frictional interfaces. Furthermore, violent frictional heating in LO_2 can lead to the ignition of tribo-elements due to burn-out phenomenon occurring in nucleate boiling, that is defined by engineering heat transfer. Under burn-out phenomenon in boiling, an extreme rise in surface temperature was experienced because a marked reduction occurred in heat transfer. For example, in boiling of LN_2, the sliding surface of Ag-10%Cu alloy (melting point 1,155 K) against Ti alloy (Ti-5Al-2.5Sn) melted due to burn-out wear during friction test [24]. The surface coating of TiN or TiO_2 had a high resistance to adhesive welding to the Ti alloy disk was able to protect from burn-out wear. The results were applied to the balance-piston system in the LH_2 turbopump of the LE-7.

CRYOGENIC TRIBOLOGICAL PROBLEMS [4, 16, 20, 23]

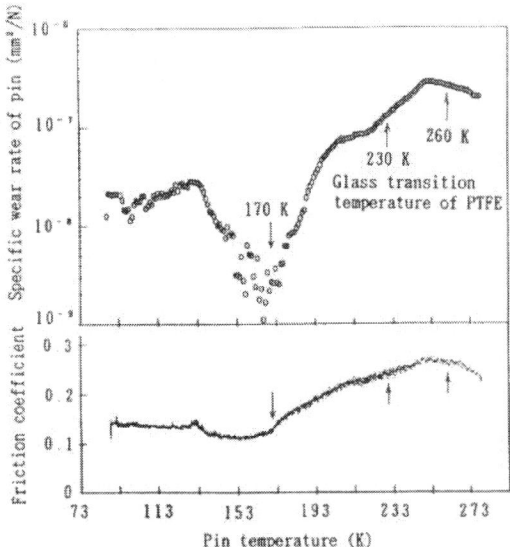

Figure 4. Friction and wear of PTFE pin against 440C disk in cryogenic GO_2 as a function of pin temperature

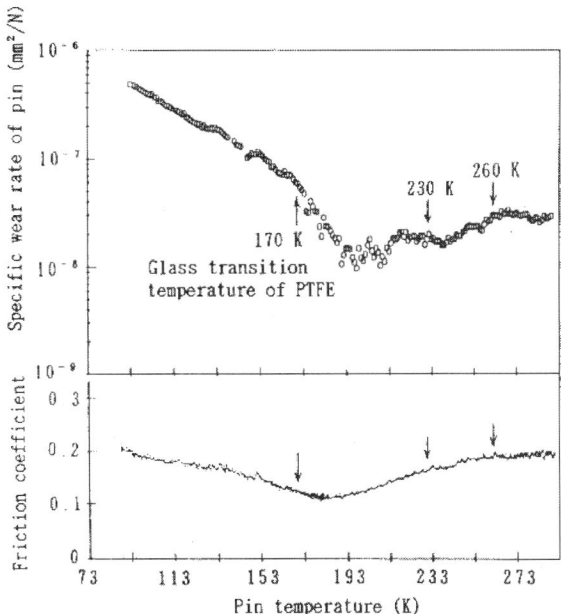

Figure 5. Friction and wear of PTFE pin against oxidized 440C disk in cryogenic GO_2 as a function of pin temperature

It is noted that the tribo-characteristics at cryogenic temperatures tend to change complexly. For example, Fig. 4 shows the change of friction and wear of a PTFE pin against a 440C steel disk in cryogenic gaseous oxygen (GO_2) as a function of pin temperature [23,25]. This figure denotes the glass transition temperature of PTFE, about 170 K, 230 K and 260 K, those are defined by relaxation of its amorphous layer in the PTFE band structure. When the frictional environment changed from the liquid phase to the gas phase at boiling, the friction coefficient increased drastically and wear began. To the glass transition temperature of 170 K (amorphous layer begins to relax), the friction coefficient remains at a low constant value, but the specific wear drastically decreased at 170 K. In an inert gaseous nitrogen (GN_2), there was not such drastically decrease in the specific wear at 170 K. After that, friction and wear begin to increase gradually up to 230 K. The increase of friction and wear above 170 K surely depends on the fact that the strength property of PTFE begins to decrease rapidly above 170 K.

However, when the surface of 440C steel was oxidized, the characteristic curve of friction and wear depended on cryogenic temperatures was changed drastically. Figure 5 shows the change of friction and wear of a PTFE pin in case of using an oxidized 440C steel disk [23,25]. At the pin temperatures above boiling point of LO_2 (90 K), the friction and wear of PTFE pin showed relatively high values as compared with that showed in Fig. 4. As the pin temperature increased from 90 K to near 170 K, the friction and wear of PTFE drastically decreased to low values. The oxidized 440C steel disk was obtained by heating in air at about 623 K for 3 hours. The surface of the oxidized 440C showed an increase of FeO/Fe_2O_3 film in comparison with Cr_2O_3 film. It is noted that the oxidization of 440C steel should result in an increase of friction and wear of PTFE. It seems that PTFE transfer film was less formed due to poor adhesion of PTFE against FeO/Fe_2O_3, and frictional condition became to be severe. Thus, it is very interesting that the friction and wear properties of PTFE changed characteristically at its glass transition temperature, depending on the oxidization of 440C steel.

For other friction tests, wear of PTFE in cryogenic GO_2 was increased as surface roughness of 440C disk was increased; however, in cryogenic GN_2, surface roughness had less effect on wear increase of PTFE. Furthermore, friction and wear of PTFE against Si_3N_4 disk was determined in cryogenic GO_2 and GN_2. In both cryogenic environments, friction coefficient was higher than that of 440C disk. It was noted that wear of PTFE in GO_2 was drastically high compared with that in GN_2. It was assumed that poor formation of PTFE transfer film on the SI_3N_4 disk resulted in an increase of friction and wear in GO_2. This result indicated that the hybrid ceramic bearing with Si_3N_4 ball showed poor self-lubrication in LO_2.

It is interesting to use ceramic material as tribo-materials in cryogenic environments. Friction and wear behavior of typical fine ceramics against 440C disk were evaluated in LO_2 and LN_2. Figure 6 and 7 show wear and friction of five kinds of the ceramic balls in comparison with those in LO_2 and LN_2[23], respectively. In all the cases of friction tests, the sliding contact surface of ceramic pin was covered by the transfer film of wear debris of 440C steel. The metallic transfer film prevented direct contact between metal and ceramic. As a result, the metal-to-metal contact should control the friction and wear behavior of the sliding pair, and the order of friction seemed to be less affected in the wear resistance of ceramic pins.

In LO_2, Al_2O_3 indicated the lowest wear rate and was followed by SiC, Si_3N_4, Sialon and ZrO_2 in order of the wear resistance. For Al_2O_3 pin, the metallic oxide film of 440C seemed to be strongly adhered onto the ceramic pin and resulted in an increase of protection of the pin wear; however, wear of the 440C disk was prolonged. For SiC, Si_3N_4 and Sialon, sliding friction in oxidized environment made the glassy formation of SiO_2 film due to tribo-chemical reaction. The hardness of SiO_2 is much less than that of ceramic substrate and resulted in an increase in the wear of ceramic pins. It was noted that the wear rate of ZrO_2 was considerably high as similar to that of self-mated 440C steels. Since ZrO_2 has the lowest hardness compared with other ceramics, the hard oxide film of 440C should increase wear of ZrO_2 pin.

To the contrary, in LN_2, Zr_2O_3 indicated the lowest wear rate and was followed by Si_3N_4, Sialon, Al_2O_3 and SiC in order of the wear resistance. The high wear of Al_2O_3 and SiC pins was seemed to be induced by lack of protective film of 440C steel due to weak adhesion to ceramic pin. It is found that the order of wear resistance of ceramics against 440C steel in LO_2 was opposed to that in LN_2 [23].

At cryogenic temperatures, it is noted that sufficient cooling and the restriction of frictional heat generation are essential to prohibit severe tribological conditions. In order to solve these cryogenic tribological problems, it is important that (1) understanding the complex characteristics of tribology at low temperatures, (2) selection of the proper solid-lubricants against the oxidation or reduction power, and (3) active cooling to remove severe frictional heat at local hot spots [4].

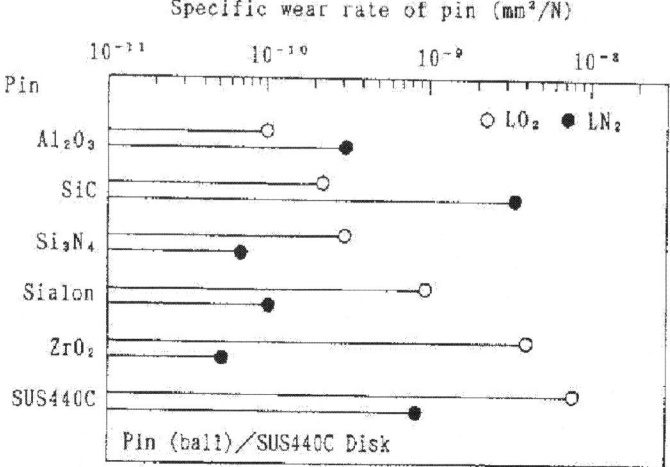

Figure 6. Wear of five kinds of the ceramic balls against 440C disk in LO_2 and LN_2

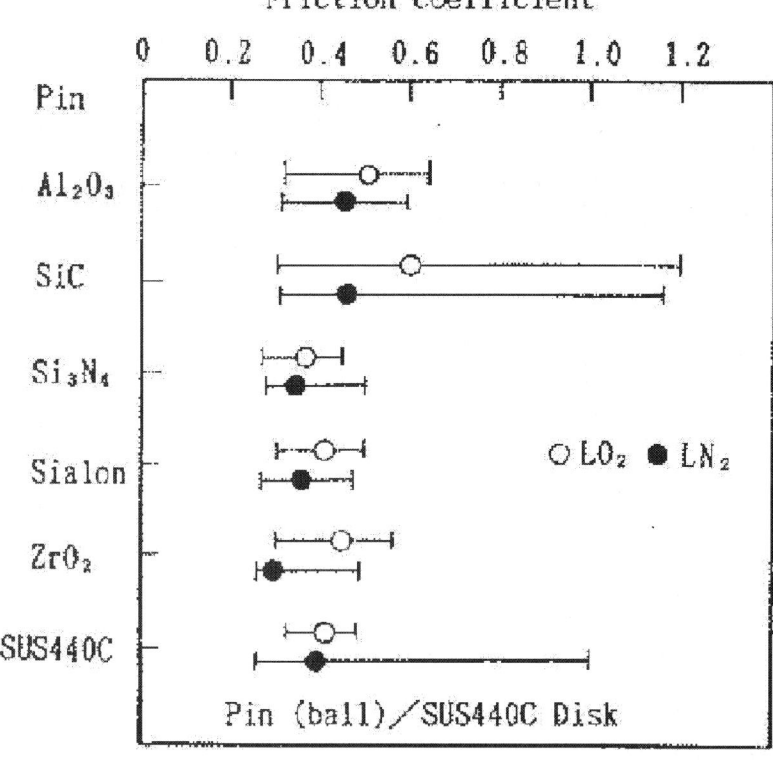

Figure 7. Friction of five kinds of the ceramic balls against 440C disk in LO_2 and LN_2

HIGH–SPEED BEARINGS

Improvement of Self–Lubrication of Retainer [16, 17, 18, 26, 27]

In the beginning of the development of the turbopump bearing for the LE-5, the bearing had used the composite PTFE retainer reinforced with glass fiber or carbon fiber. The bearing tested in LH_2 by using a bearing tester showed that the glass fiber-reinforced PTFE retainer (24 wt.% glass fiber and additive) could demonstrate stable bearing-torque performance as compared with that of the carbon fiber-reinforced retainer (15 wt.% carbon fiber). From inspection of the ball-pocket surface of the carbon fiber-reinforced retainer, it was found that pile-up of the wear debris of carbon fiber might reduce supply of PTFE transfer film to ball surface. As a result, the LH_2 turbopump bearing selected the glass fiber-reinforced PTFE retainer; however, the real turbopump test showed severe wear of the retainer when the turbopump was operated under poor cooling conditions. This fact indicated low wear resistance of the glass fiber-reinforced PTFE retainer under severe operation of turbopump [16,17].

For the rocket-turbopump bearings, a laminated glass cloth with PTFE binder (laminated glass cloth of 45 wt.% and PTFE of 55 wt.%) was currently used because of its great strength to protect against dangerous retainer rupture [4,17]. This retainer showed poor self-lubrication resulting from abrasion by glass cloth layers exposed on the ball-pocket surface. During the development of the LH_2 turbopumps for the LE-5, the bearing showed unstable high-temperature rise and poor lubrication was observed, resulting in severe wear of the balls. In case of the reusable turbopumps used in the SSME, the bearings similarly experienced a serious wear problem [6]. In order to improve the self-lubricating performance of the retainer, special surface treatment of the retainer was developed [12,18]. The abrasive retainer surface with the exposed glass cloth was chemically etched with hydrofluoric acid (HF) to a depth of 0.10-0.15 mm. Following this treatment, smooth surface for the retainer was obtained. The sliding friction and wear between the ball and ball-pocket surface was reduced, resulting in a sufficient supply of PTFE transfer film from the retainer to the rolling balls.

For the HF etched retainer tested in LH_2, detailed examination of the transfer film on the sound ball surface with hardly any wear was conducted by electron probe microanalysis (EPMA) [12]. The result indicated that F of PTFE of the retainer strongly depended on the Ca concentration on the map and resulted in the tribo-chemical formation of CaF_2 transfer film. The reacted oxide material (49 wt.% of glass fiber) consisted mainly of an oxide of Ca (CaO) remained on the HF etched retainer surface. Therefore, it seems that the formation of CaF_2 transfer film was conducted by tribo-chemical reaction between F of PTFE and CaO remained on the retainer surface in chemical reduction environment in LH_2.

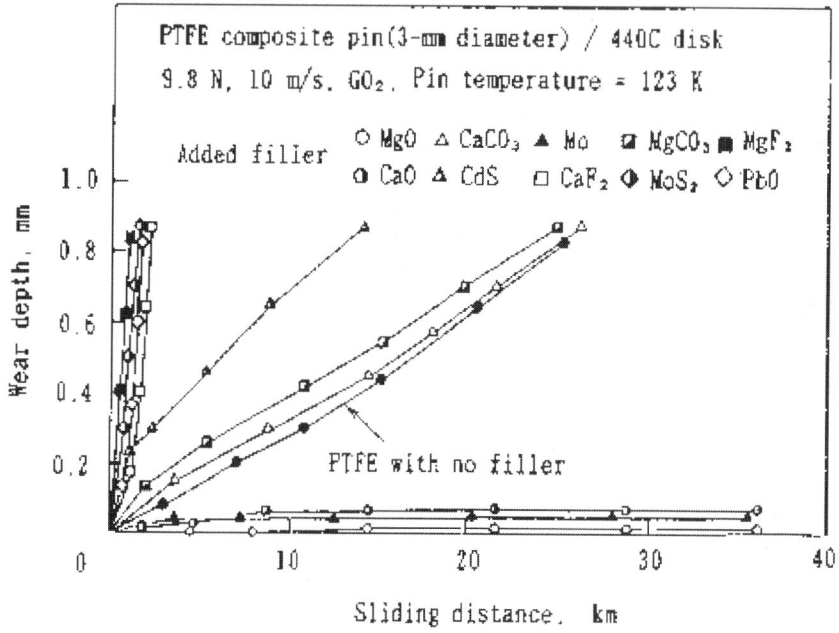

Figure 8. Wear of PTFE composite pins with various fillers against 440C disk in cryogenic GO_2 (123 K) under high-sliding speed (10 m/s)

In order to determine the effect of tribo-chemical formation of CaF_2 in transfer film, additional friction tests were conducted. Figure 8 shows the wear of PTFE composite pins with 15 wt.% of various fillers against the 440C disk in cryogenic oxygen gas (GO_2, 123 K) under a high sliding speed (10 m/s) [15]. The PTFE composites with CaO and MgO fillers showed excellent wear resistance (progression of the pin-wear was stopped) due to the formation of good transfer film even in both cryogenic GO_2 and GN_2(123 K). It seems that alkali-earth-metals such as Ca and Mg were able to react easily with F by severe dry sliding friction and resulted in the formation of CaF_2 and MgF_2 within the transfer film [4]. The tribo-chemical formation of CaF_2 and MgF_2 might enhance adhesion of transfer film. When CaF_2 and MgF_2 added as fillers to PTFE, there was no tribo-chemical reaction, resulting in poor wear resistance. Furthermore, oxidation of the Mo filler in GO_2 seemed to be extremely effective except in GN_2.

Development of Elliptical Ball–Pockets of Retainer [13, 14, 26]

During testing of the LH_2 turbopump for the LE-7, the conventional bearings using a retainer with circular pockets showed a significant temperature rise under high shaft vibration. Since high shaft vibration increases the radial load applied to the bearings, ball excursion occurring in the ball pockets of the retainer due to ball-speed-variation (BSV) becomes significantly large. Figure 9 shows the ball excursion due to the BSV *vs.* the radial load for the 40-mm-bore bearing at a speed of 42,000 rpm [13]. The ball excursion tends to increase with increasing of the radial load. At a radial load of about 1.5 times thrust load, the ball excursion reaches a maximum value. When the pocket clearance of the retainer is smaller than the maximum ball excursion, severe contact occurs between the ball and the retainer pocket.

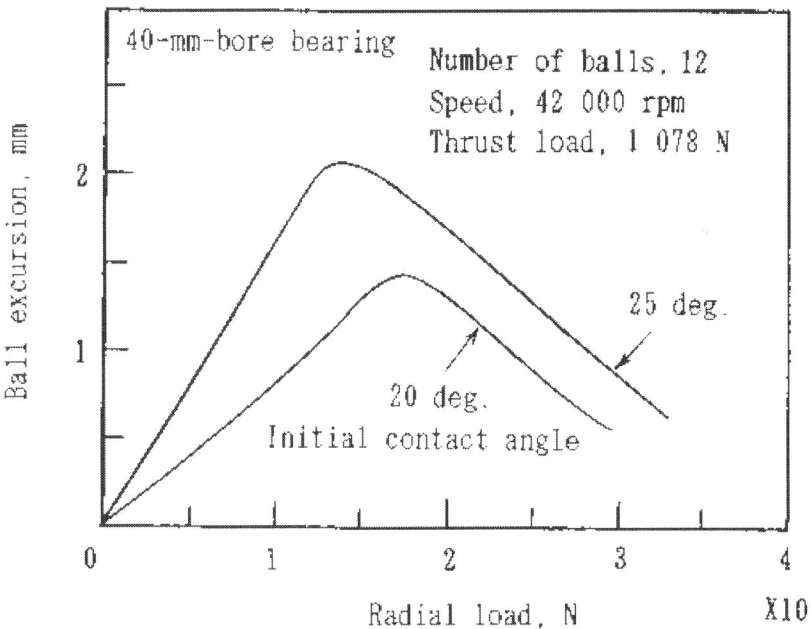

Figure 9. Ball excursion due to BSV *vs.* radial load for LE-7 LH_2 bearing at 42,000 rpm (40-mm-bore bearing)

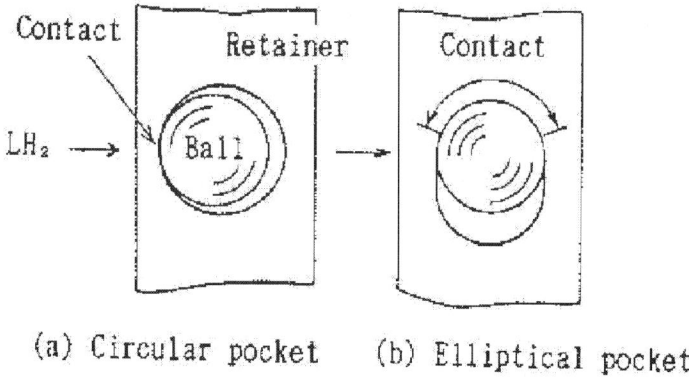

Ball pocket	Ball-pocket clearance, mm
Circular	0.9
Elliptical	1.8 (Circumferential) 0.3 (Axial)

Figure 10. Circular and elliptical pockets of retainer and ball pocket clearances for 40-mm-bore bearing

For the 40-mm-bore bearing, a retainer having elliptical pockets with a large pocket clearance was developed. As shown in Fig. 10, this retainer with elliptical pockets is able to allow maximum ball excursion due to BSV in the circumferential direction and to stabilize wobbling of the retainer due to a narrow clearance in the axial direction [13]. The pocket clearance of 1.8 mm was twice as large as that of the conventional circular pocket. Consequently, the LE-7 turbopump bearings with the elliptical-pocket retainer exhibited excellent performance by reducing severe frictional heating and high wear of bearing components at a high-speed level of 50,000 rpm (2 million DN). Basic study of the elliptical pocket of the retainer was conducted in the development of the LE-5 turbopump bearing [12,17].

During the development of the LE-7A, the LH_2 turbopump experienced severe operation with high vibration of the rotating shaft. As a result, high vibration of the rotating heavy turbine-disk increased radial load at the turbine-side bearings (40-mm bore) and broke the retainer due to large BSV [26]. It was considered that the ball-retainer contact force due to BSV bent the retainer and hoop stress occurred on the retainer inside, resulting in fracture of the thin (weak) web section of the ball pocket. In order to gain high reliability of the LH_2 turbopump, the retainer using elliptical ball pocket was improved by increasing the pocket clearance to 2.2 mm.

HIGH–SPEED BEARINGS

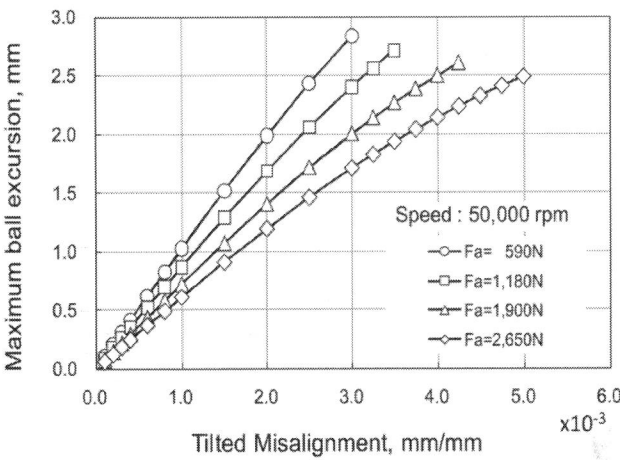

Figure 11. Maximum ball excursion vs. tilted misalignment under various thrust loads at 50,000 rpm (40-mm-bore bearing)

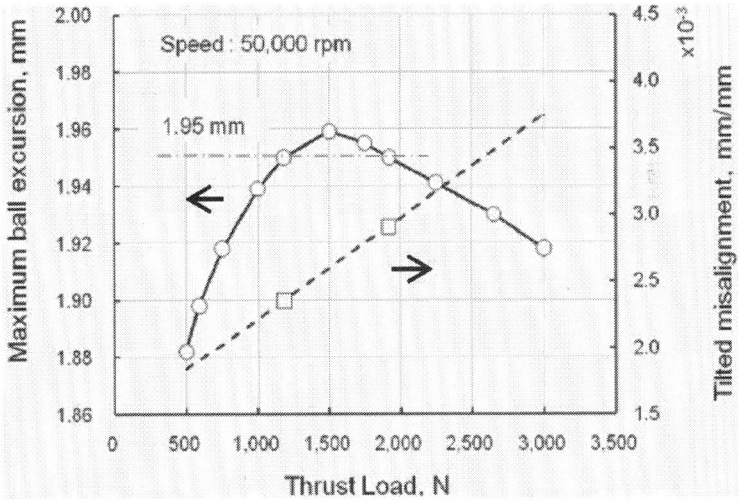

Figure 12. Maximum ball excursion and tilted misalignment *vs.* thrust load at 50,000 rpm (40-mm-bore bearing)

Such BSV was also caused by inclination of the outer race to the shaft (tilted misalignment). The effect of tilted misalignment in a level of 1.9-3.5 x 10^{-3} mm/mm on the tribo-characteristics of 40-mm-bore ball bearing was determined. The bearing used a retainer having various elliptical ball pockets to restrain the ball-retainer contact due to high BSV. The elliptical ball pocket changed the pocket clearance (1.75mm, 1.95 mm and 2.15 mm). Figure 11 shows the relationship of the tilted misalignment and the maximum ball excursion under various thrust loads at a speed of 50,000 rpm [26]. It is understood that maximum ball excursion increased with an enlargement of tilted misalignment.

Figure 12 shows the relationship of the maximum ball excursion and the tilted misalignment *vs.* the thrust load at a speed of 50,000 rpm [26]. The relationship of the maximum ball excursion *vs.* the thrust load was calculated by assuming that the tilted misalignment linearly increased with an increase of the thrust load. As the thrust load increased, the calculated maximum ball excursion tended to increase in a parabolic pattern. It was found that, in case of the pocket clearance of 1.95 mm, ball-retainer contact due to ball excursion possibly occurred within a limited range of thrust loads, resulting in high increase of bearing torque and bearing temperature.

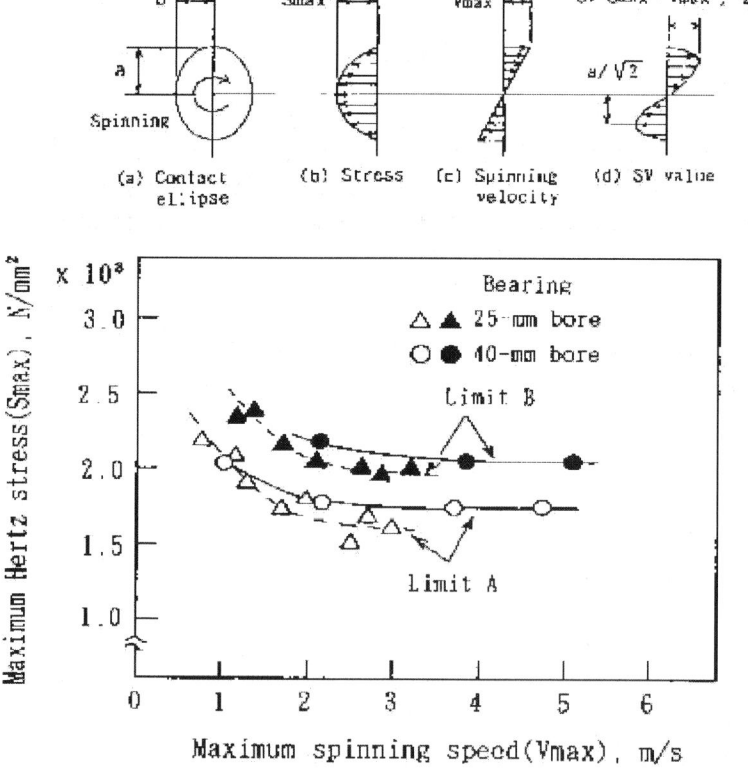

Figure 13. Load capacity of transfer film under inner race ball-spinning in LH_2

HIGH–SPEED BEARINGS

Performance of Lh₂ Bearing [12, 13]

Performance of self-lubricating bearing coated with PTFE or MoS_2 films was evaluated for the LH_2 turbopump bearing of the LE-5. The PTFE and MoS_2 films were coated with rf-sputtering. Bearing test was conducted for about 2 hours at a speed of 50,000 rpm in LH_2. Frictional heating was estimated from the temperature rise of cooling flow through the test bearing [12]. The coated films are hoped to induce smooth running in the initial operation when the amount of the PTFE transfer film is insufficient. The high self-lubricating performance and durability were experimentally confirmed with the PTFE coated bearing indicating frictional heating of 170-250 W. For the MoS_2 coated bearing, the frictional heating was 250-330 W and relatively high. The retainer of the PTFE coated bearing showed less ball-pocket wear than that of the MoS_2 coated bearing.

For high-speed bearings, since the bearing was under the outer-race ball control at high speed, the transfer film of the inner raceway was damaged due to the spinning of the ball. In order to evaluate stable operating condition without bearing damage, the load capacity of the transfer film under inner race ball-spinning in LH_2 was determined as shown in Fig. 13[13]. This figure shows the critical load capacity, that is, maximum Herze stress ($Smax$) vs. maximum spinning speed ($Vmax$). Under high thrust loads, an increasing of the bearing torque and bearing temperature (at limit A) was determined by the bearing tester which could measure the bearing torque in LH_2. The film local rupture (at limit B) was also defined by the electrical resistance monitoring between the inner race and outer race. Up to a $Vmax$ of 5 m/s at 50,000 rpm, the transfer film was able to sustain a $Smax$ up to 2 GPa. It was determined that the load capacity of the transfer film depended more on $Smax$ than on $Vmax$. So, in order to increase durability of the bearing, it is important to limit the stress level to a $Smax$ of 2 GPa to prevent transfer-film rupture and sufficiently to cool the frictional heat due to high $Vmax$.

Durability of Lo₂ Bearing [15]

It is noted that violent frictional heating in LO_2 can lead to the ignition of tribo-elements due to burn-out phenomenon. Burn out is overheat occurring in a transition from nucleate boiling to film boiling at critical heat flux that is defined by engineering heat transfer. For the LO_2 turbopump bearings (32-mm and 45-mm bore) of the LE-7, the durability and fatigue life were evaluated by applying heavy radial loads at a speed of 20,000 rpm in LO_2 or LN_2. During testing, the bearing-cartridge-acceleration (BCA), i.e., Gpk (peak value) and $Grms$ (rot-mean-square value), was monitored to detect bearing damage. Testing in LO_2 for about 2.2 hours under a system radial load of 5,880 N showed that excellent lubricating conditions without abnormal BCA were obtained for all bearings.

Durability test in LN_2 (to keep safety in the experience) under a heavy system radial load of 11,760 N was conducted at a speed of 20,000 rpm for about 5.1 hours [15]. The result detected that the fatigue life of the bearing was about the same as the calculated B_{10} fatigue life. The bearings were operated at steady conditions for 5.1 hours with 20 start-stops. For BCA on bearings A/B, *Gpk* and *Grms* on the chart were abnormally separated from each other in a pattern of abnormal BCA showing an increase of surface roughness due to an occurrence of slight flaking. Then, at a total test time of 3.8 hours, the loaded and unloaded BCA abnormally began to increase concomitantly. Examination of tested bearing B indicted that slight flaking with very shallow depth (about 8.5 µm) was observed on the inner raceway.

Evaluation of Turbopump Bearings [14]

The durability of the bearings of the LO_2/LH_2 turbopumps used in the firing tests of the LE-7 was evaluated based of findings of wear inspection and X-ray photoelectron spectroscopic (XPS) analysis of PTFE transfer film. Inspection of the turbopump bearings used in the engine firing tests is essential for evaluation of their durability under engine operation.

a. *Bearing wear*

After the engine firing test, surface profiles of the raceways of the LH_2 turbopump bearings was evaluated [14]. The engine test was conducted for a total time of 31.4 minutes with 20 engine start-stops. The surface profiles included the thickness (1µm) of the initial film coatings of sputtered PTFE film. It is obvious that the wear scars on the raceways of all bearings were flat and spin wear was not observed despite conditions of higher ball spinning on the inner raceway. For the retainer with elliptical pockets, the wear depths in the pockets were smaller than the depth (0.10-0.15 mm) of chemical etching of the glass cloth. The PTFE layer without the abrasive glass cloth sufficiently remained at the bottom of the pocket wear scar.

To contrary, the all inner raceways of the LO_2 turbopump bearings showed typical spin wear with light oxidative wear [14]. These turbopump bearings tested for a total time of 34.6 minutes with 23 engine start-stops. The surface profiles included the thickness of the initial film coatings of sputtered PTFE film (1 μm) on Ion-plated Au film (0.4 μm). The wear depths of raceways seemed to be relatively high; however, smooth surface roughness demonstrated mild wear without severe adhesion due to metal-to-metal. For bearing D that was affected by turbine whirling with radial overload, heavy spin wear with a wear depth of 7 μm was generated on the inner raceway. Furthermore, slight flaking was observed on the inner and outer raceways. This flaking was characterized by a very shallow depth and by fractures on the surface.

For the retainer with conventional circular pockets, the wear depths in the pockets were relatively light compared with those of the LH_2 bearing. The contact area in the retainer pocket and on the ball surfaces was blackened by the thermally degraded transfer film. The degradation of the transfer film seemed to occur at a temperature above about 500 K. This was confirmed by a heating test of the retainer. These facts indicated that the transfer film was severely heated even in cryogenic fluid and the LO_2 turbopump bearings were operated under poor cooling conditions. Thus, to increase the durability of the bearings, it is apparent that sufficient cooling is essential.

a. *XPS analysis of transfer films*

In order to evaluate the excellent lubricating conditions without severe wear, XPS depth analysis of a transfer film on a ball used in the LH_2 turbopump bearing of the LE-7 was conducted. Inspected ball that showed excellent wear condition was from the turbine-side bearing tested for 31.4 minutes in engine tests. The XPS depth analysis with an etching depth of 30 nm (SiO_2 rate) indicated that F(1s) and Fe(2p) spectra show the significant formation of thick CaF_2 and FeF_2 film as shown in Fig. 14 [4]. It seemed that, due to the reduction power of LH_2, the reacted CaO (remained on the retainer surface chemically etched with HF) was tribo-chemically changed to CaF_2 with the F of PTFE retainer during bearing operation. In addition, due to removing of native oxide film by the LH_2 reducing power, a FeF_2 film was formed by a chemical reaction between the F of PTFE retainer and the Fe of 440C steel. It is noted that the formation of FeF_2 film at the stressed contact area resulted in demonstrating high resistance to metal-to-metal adhesion and in leading to less wear [27].

Thus, the LH_2 turbopump bearings used in the engine firing tests demonstrated excellent performance due to the formation of thick CaF_2 and FeF_2 film. The tribo-chemical formation of CaF_2/FeF_2 film possibly reduced wear at frictional interfaces within the bearings used in LH_2. The basic tribo-chemical reaction was determined as follows [4]:

$$(-CF_2-)_n + CaO + Fe \rightarrow (-CF_2-CO-)_n + CaF_2 + FeF_2$$

(1)

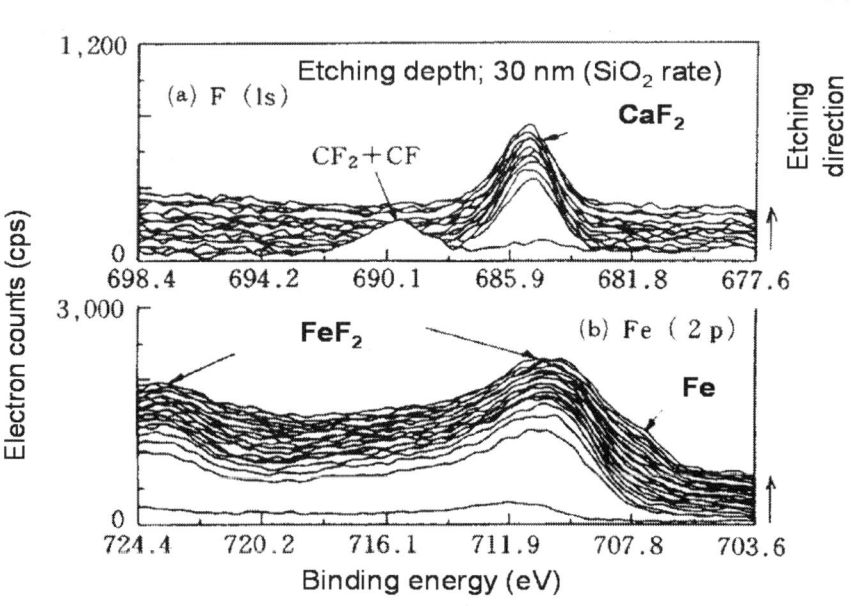

Figure 14. XPS depth analysis of ball for LH_2 turbopump bearing (turbine side)

On the contrary, for the LO_2 turbopump bearings of the LE-7, the inspected ball was from the turbine-side bearing that was tested for 34.6 minutes in engine tests and showed heavy spin wear. Figure 15 shows the XPS depth analysis with an etching depth of 30 nm (SiO_2 rate) for the worn ball due to spin wear. It indicated that the oxidization power of LO_2 prohibited the tribo-chemical formation of CaF_2/FeF_2 transfer film. This bearing was operated under poor cooling conditions, so that the bearing wear was relatively increased and shallow flaking was formed on the raceways. From the F spectrum, it was shown that very thin $PTFE/CaF_2$ transfer film was formed compared with the thick $PTFE/CaF_2$ transfer film in the LH_2 bearing. Furthermore, from the Fe spectrum, formation of Fe_2O_3 oxide film was typically shown. Fe_2O_3 oxide film was apt to form at elevated temperature, so that the oxidative mild wear in the bearing was increased due to poor cooling conditions in LO_2 [5]. As mention later (in 6.1.1), for the bearing tested under sufficient cooling condition, the intense formation of Cr_2O_3 film without Fe_2O_3 film was found beneath an extremely thin PTFE film, resulting in high resistance to metal-to-metal adhesion and in a decrease of the bearing wear [28].

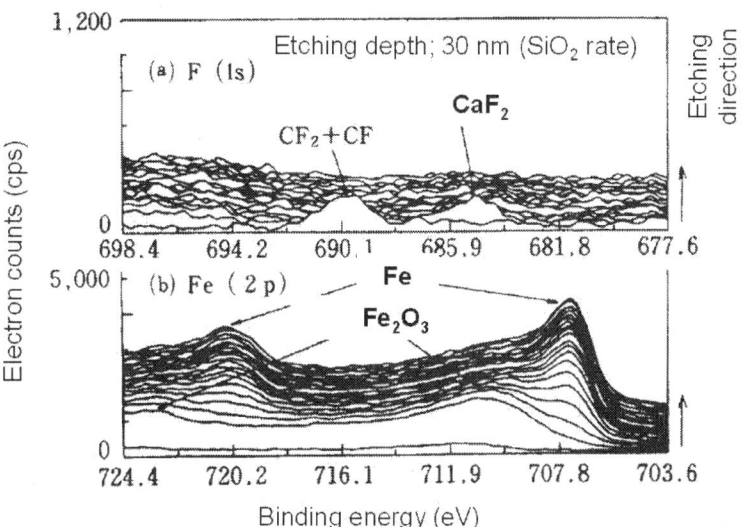

Figure 15. XPS depth analysis of ball for LO_2 turbopump bearing (turbine side)

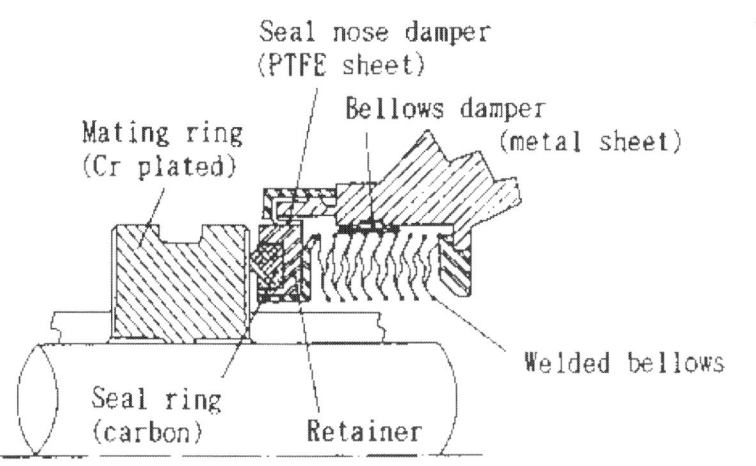

Figure 16. Face-contact mechanical seal for LH_2 turbopump of LE-5

TURBOPUMP SHAFT SEALS

Mechanical Seal [29-34]
For the LE-5 turbopumps operating under the gas generator cycle, the contact-type mechanical seal was able to use for the propellant seals because the pump and turbine pressures were relatively low. Specially, for the LH_2 turbopump, a high-speed mechanical seal was required to withstand high rubbing speed (113 m/s) at a speed of 50,000 rpm in LH_2. Figure 16 shows the face-contact mechanical seal with a seal diameter of 43.2 mm developed for the LH_2 turbopump of the LE-5 [29,30]. In order to reduce seal leakage of LH_2, it has a modified seal nose that could reduce the seal face distortion and control the direction of its distortion (to contact at outside of the seal face) under low temperature and high pressure. Furthermore, a modified vibration damper made of PTFE sheets is attached around the seal nose to prevent fluttering during rapid start or stop of the turbopump.

When the closing force to contact seal faces is increased to make seal leakage smaller, wear rate of the seal faces is increased due to the poor lubrication of LH_2. If the closing force is set to be smaller than the fluid opening force separating seal face each other, the leakage is considered to be quite large because of the extremely low viscosity and density of LH_2. Therefore, to obtain the stable seal performance and the long wear life, it is important that the proper balance between the closing force and the opening force is retained.

Critical value of the seal balance ratio that obtained stable seal performance and reduce wear of the seal faces was experimentally and analytically evaluated [32,33]. In this study, the experimental and analytical study on the friction power loss and seal performance was conducted. It was indicated that the friction power loss fell to a small value after the seal faces were sufficiently run-in. The seal balance ratio *[B]* that stabilized seal performance was in a range of 0.77-0.82. The seal balance ratio*[B]* is determined by the following equation;

$$[B] = B + Fsp/(As\Delta P)$$

(2)

where, *B* is the fluid balance ratio, *Fsp* is the spring force of bellows, *As* is the seal area and *ΔP* is the seal pressure. *[B]* is determined by the initial spring force of the bellows.

When the seal balance ratio was below 0.77, the leakage was apt to increase due to lack of the closing force. In this case, the critical balance ratio *[B]c* that gains stable seal performance showing small leakage was 0.77. To contrast, its balance ratio above 0.82 increased wear of the seal face by rise of the closing force. This high value of critical balance ratio was due to large opening force that could be explained with leakage flow model, assuming the phase change of leakage (from liquid phase to gas-liquid phase and gas phase) due to viscous frictional heating at high rubbing speed. In this phase change model, a state change of gas was assumed to be irreversibly adiabatic and a curve of gas expansion expressed by the following equation;

$$Pv^m = \text{constant}$$

(3)

where, P is the pressure, v is the specific volume and m is the ausfluss exponent. As m decreases with the temperature rise of gas due to viscous friction, the pressure of leakage flow increases particularly in the gas region within gas-liquid phase, and it resulted in the increase of the opening force. The analysis of phase change model of leakage was conducted using the flow and energy equations of liquid and gas leakages.

Figure 17 shows the calculated and experimental results of the relationship between the seal clearance and the opening force ratio *[K]* at a speed of 50,000 rpm in LH$_2$. The opening force ratio *[K]* is expressed by the following equation;

$$[K] = Fo/(As\Delta P)$$

(4)

where, Fo is the opening force. It was also shown that the opening force within seal clearance increases linearly as the seal clearance decreases. After the seal faces were sufficiently run-in and the seal clearance was maintained in an average of 0.6 µm, the opening force ratio *[K]* approaches the critical balance ratio *[B]c* (= 0.77) that showed critical seal performance. As a result, the difference of *[K]* and *[B]c* was decreased and it resulted in the reduction of the load on the seal faces. The frictional loss power was decreased to a small value, resulting in a restrain of wear rate of seal faces. If the seal clearance increases, the leakage becomes large; however, the load on the seal faces is increased with the decrease of the opening force and the seal clearance would become small enough to reduce leakage. Furthermore, the starting torque and static seal performance were markedly affected by the change of the seal face distortion due to wear [31].

Durability of the mechanical seal was evaluated by the long-run test [29]. The long-run test was conducted at a speed of 50,000 rpm with a seal pressure of 1.37 MPaG for 83 minutes. The experimental results showed that the leakage gradually increased until total test time was 50 minutes. During its step, wear of the seal faces was running-in, then the leakage was stabilized. It is noted that an extremely small LH_2 leakage (8-19 cc/min) was kept during test. The seal after the durability test indicated an excellent condition that maximum wear of carbon-ring was 8 μm.

Temperature on the rubbing seal faces was estimated from the reduction rate of the hardness of hard Cr plating on the rotating mating ring [34]. The estimated temperature of rubbing seal face was possibly reached to be about 773 K at a rubbing speed of 113 m/s in LH_2. In an initial stage of running-in, extremely high temperature of the seal faces caused thermal cracks in wear surface of the Cr plating, so that it is necessary to cool the contacting seal faces sufficiently. When the cooling of the sealing unit is insufficient, the surface of the carbon seal ring showed abnormal wear. Furthermore, the Cr plating showed better wear results than the tungsten carbide (WC) coating, because the Cr plating easily forms thin transfer films of graphite contained in the carbon. In the case of the WC coating, the transfer film of graphite was hardly formed in LH_2, resulting in an occurrence of severe seal wear.

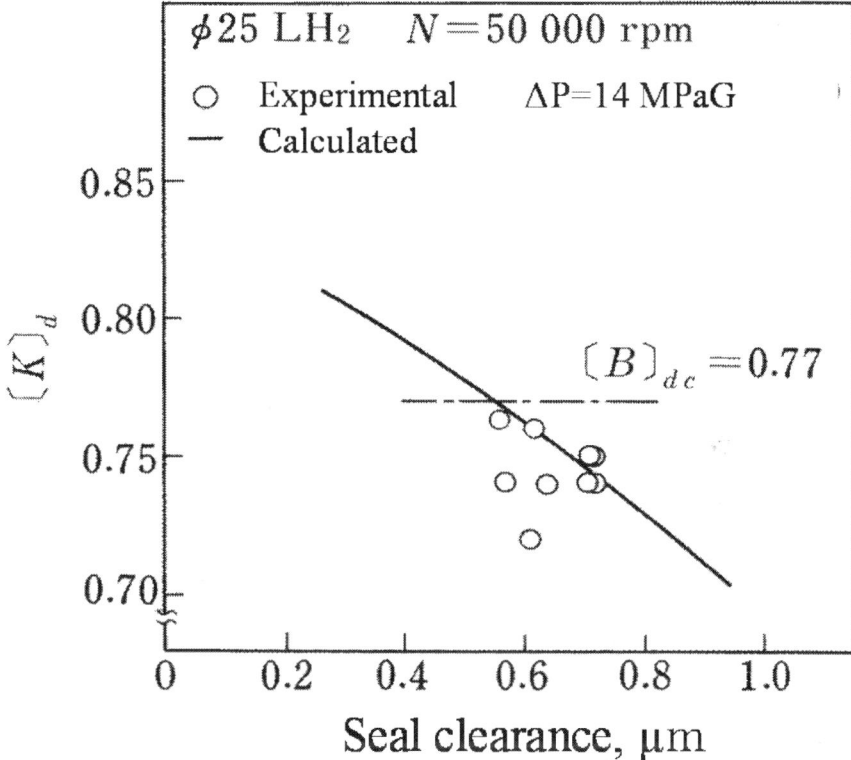

Figure 17. Opening force ratio $[K]$ vs. seal clearance at 50,000 rpm in LH_2

Floating Ring Seal [22, 29, 35, 37]

A floating-ring seal is a type of no-contact annular seal without a rubbing seal surface. It has a simple structure and is able to seal high-pressure fluids, restraining leakage through a small clearance (gap) between the seal ring and the runner. Its gap is in an order of several dozens of μm. The seal ring is free to move in the radial direction, and thus severe contact with the rotating runner can be prevented. Leakage of floating-ring seal is much larger than that of face-contact mechanical seal, but the floating-ring seal shows a high resistance to pressure and a high reliability when used as high-pressure seal. A multi-seal system consisting of several seal rings arranged in series is employed for the high-pressure turbopumps. The floating-ring seals were developed and used in the LE-5 and LE-7.

Figure 18 shows the floating-ring seal with a seal diameter of 50 mm developed for the LO_2 turbopump of the LE-7 [22,35]. The carbon seal ring is enclosed with a retainer of the same material as the seal runner.

Since the retainer contracts thermally nearly as much as the seal runner at low temperature, the seal gap hardly changes. The seal gap was 50-60μm. When the seal pressure increases, the floating ring is pressed against the secondary seal by the fluid force and its movement in the radial direction is restrained. In order to smooth the radial movement of the floating ring, on the secondary seal of the housing, the PTFE film was coated for the LO_2 seal and the MoS_2 film was coated for the turbine gas seal (to seal the low temperature GH_2). For the GH_2 leakage of the floating-ring seal used in the turbine gas seal, leakage rate calculated by the quasi-one-dimensional compressible flow equation agreed quite well with experimental value.

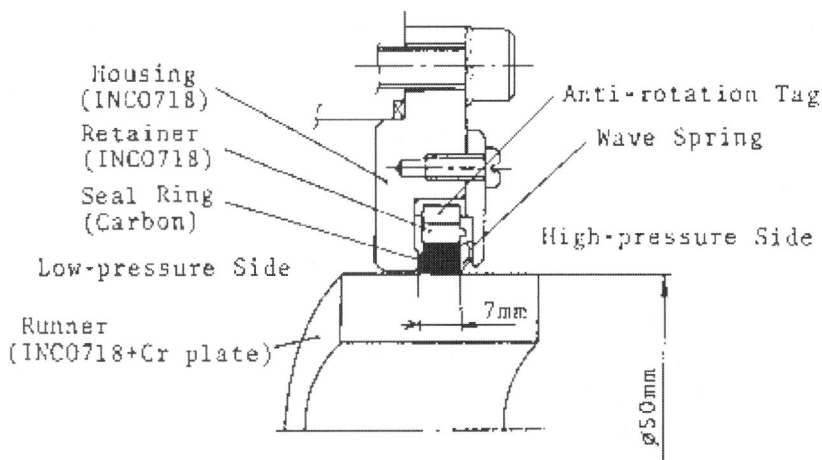

Figure 18. Floating ring seal for LO_2 turbopump of LE-7

The leakage from the floating-ring seal for the LH_2 and LO_2 seal can be calculated from the equation of the incompressible fluid flow in the rotating double cylinders when the leakage is liquid phase flow and the mass flow flux (mass flow/seal area in the flow direction) is large [29,35]. When the seal gap is narrow and the seal pressure is low, the mass flow flux of leakage is reduced, and vaporization of leakage occurred by viscous frictional heating and pressure drop changes liquid phase flow to gas-liquid phase flow (two-phase flow).

Comparison between the experimental and calculated leakage of LH_2 was evaluated by the mass flow flux of leakage for the floating-ring seal with one seal ring or two seal rings [29]. In this study, the LH_2seal with a seal diameter of 32 mm and various seal gap of 30-86 μm was tested at rotating speeds to 50,000 rpm. It is shown that the leakage of LH_2 is less than the calculated value from incompressible fluid flow equation because the leakage is changed to be tow-phase flow. When the mass flow flux is large, most of leakage flows out in liquid phase. This means that there is not sufficient time to vaporize the leakage to be tow-phase flow within the seal gap.

A flow visualization study of floating-ring seal was conducted to identify the two-phase flow area induced by viscous frictional heating and pressure drop [36]. In order to visualize the two-phase flow in seal gap, the floating ring made of transparent hard plastic (polycarbonate) was tested in a seal fluid of LN_2. It was confirmed that the two-phase flow seemed to be homogeneous mixture of liquid and vapor flow and the two-phase flow area increases with increasing rotational speed and decreases leakage flow rate. When the two-phase flow area was fully prolonged within the seal gap, the leakage rate contrary increased with instability because the inlet flow resistance at the high-pressure side of the seal ring was reduced by two-phase flow.

Segmented Seal [22, 35, 37, 38]

Contact-type segmented seal were used in the GHe purge seals and the low pressured turbine gas seals. The GHe purge seal used in the LO_2 turbopump of the LE-7 is shown in Fig. 19 [22]. Segmented seal has a carbon seal ring cut into three segments. The segmented annular seal ring is pressed on the seal runner with a coil spring and maintains high purge-pressure of GHe as a barrier gas. Wear of the carbon seal ring is reduced by using the shrouded Rayleigh step lift-pads to increase the opening force within the seal clearance. As the rubbing speed increases, the opening force in the Rayleigh step increases, so that the rubbing speed is increased by enlarging the seal diameter using a T-type runner.

Relationship between the purge pressure and the leakage rate of GHe purge seal was evaluated at a steady speed of 20,000 rpm [22]. When the purge pressure is low, the seal face is kept to be non-contact because the Rayleigh step increases the seal opening force. As the purge pressure is set to be high, the seal face condition is changed from the non-contact state to the contact state, it resulted that the dynamic leakage almost equals that of the resting state. Furthermore, for the GHe purge seal combined with the LO_2 floating-ring seal, the environmental temperature around the GHe purge seal was equal to that of LO_2 leakage, so that the carbon seal ring showed severe wear with an appearance of worn-out of the Rayleigh step.

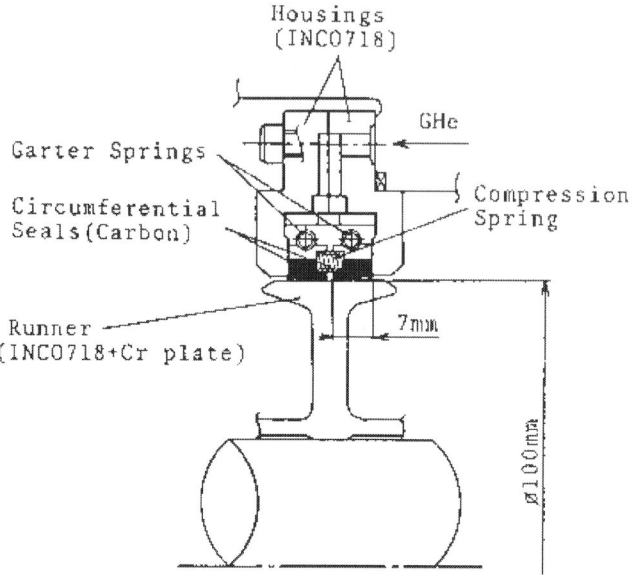

Figure 19. GHe purge seal for LO_2 turbopump of LE-7

Figure 20. Comparison of wear of MoS_2 coated and uncoated seal surfaces

Change of the friction and wear of the carbon pin as a function of the pin temperature was determined in the cryogenic GHe environment [23]. Friction test was conducted against the Cr-plated steel disk at a sliding speed of 12 m/s and load of 9.8 N. When the pin temperature is below the solidification temperature of CO_2 (216 K), it is noted that lubricating property of the carbon pin suddenly disappeared and friction and wear became intensive. When absorbed CO_2 gas was changed to be solid phase, lubricity of carbon was lost. This phenomenon resembles that when phase of moisture is transfer to solid phase (ice) below 273 K, lubricity decreases; be well known. From this fact, it seemed that severe wear of the GHe purge seal was generated because the environmental temperature around the seal was lower than 216 K. Spray MoS_2 coating on the carbon seal face was drastically able to prohibit progression of wear of the carbon seal ring at low temperature, as shown in Fig. 20.

After a total operating time of 29 minutes for the engine firing test, the GHe purge seal used in the LE-7 indicated that the seal surfaces coated by MoS_2 were found to be in excellent condition and wear depth of the carbon seal ring was about 7 μm. It assumes that high opening force produced by the Rayleigh step was kept by prohibit of wear of the Rayleigh step and the GHe purge seal was operated under conditions of nearly no load on the seal surfaces due to balance between the opening and closing forces.

ADVANCED BEARINGS AND SHAFT SEALS

Future space transport systems require reusable launch vehicles to reduce launch cost and to increase efficiency. The durability of reusable turbopump bearings must be greater than that of currently available (expendable) turbopumps. For the improved high-pressure LO_2 turbopump of the SSME that reduced serious wear of the all-steel bearing, the hybrid ceramic bearing with Si_3N_4 balls was developed and accomplished the required life of 7.5 hours. In this case, to improve self-lubrication of the abrasive retainer made of glass cloth-reinforced PTFE, a new type of the retainer that had PTFE/bronze-powder insert fitted on the ball pocket was developed [7].

It is noted that, at high speeds, the hybrid ceramic bearing that consists of hard, light weight ceramic balls as well as steel rings shows a lower centrifugal force on the ceramic ball. The centrifugal force of the Si_3N_4 ball makes about 60 % lighter than that of the 440C steel ball. This leads to a reduction of bearing load and a smaller contact area with a lower spinning speed, resulting in a low level of heat generation due to ball spin. Additionally, good tribological combinations of the ceramic balls against the steel rings result in a decrease in bearing wear and in instances of seizure, even under insufficient lubricating conditions. Thus, the hybrid ceramic bearing enables higher speed operation rather than the all-steel bearing.

On the other hand, advanced rocket engines that are characterized by high performance (light weight) and high durability (long life) are required today. Ultra-high speed turbopump having a rotational speed level of 100,000 rpm needs to make engine smaller and lighter. Hybrid ceramic bearing is suitable to ultra-high speed turbopump because of lower centrifugal force. In recent years, these advanced research and development on the hybrid ceramic bearing are actively underway.

Single–Guided Bearing [27, 28, 39]

In order to increase the durability of self-lubricated bearing, it is apparent that sufficient cooling and restriction of the frictional heat generation in the bearing are essential. Its notification is experimentally identified by a series of studies on the turbopump bearing. In order to improve internal coolant flow through the bearing and to reduce bearing frictional torque, a new type of bearing having a single-guided retainer was developed. Figure 21 shows the 25-mm-bore bearing having a single-guided retainer with elliptical ball pockets [39]. The single-guided retainer is guided only by one side of the outer-ring bore (land) to reduce land friction and to increase the cooling ability within the bearing. However, reduce of retainer guiding is apt to generate unstable wobbling at high speed, so that the elliptical ball pockets with narrow axial clearance is needed to reduce wobbling of the retainer. For the elliptical ball pocket of the single-guided retainer, its circumferential clearance of 1.3 mm was twice as large as that of the conventional circular pocket to reduce ball-to-pocket interaction under high BSV. Furthermore, the axial clearance of 0.1 mm was narrow to stabilize wobbling of the single-guided retainer at high speeds.

Self-lubricating performance, bearing wear and transfer film of two-types of the single-guided bearing, i.e., a hybrid ceramic bearing with Si_3N_4 and all-steel bearing, was evaluated under high thrust loads at speeds up to 50,000 rpm in LH_2, LO_2 and LN_2 [27,39]. Furthermore, to evaluate the durability of the single-guided bearing for long-life bearing, the all steel bearing was tested for total operation times up to 11.7 hours at a speed of 50,000 rpm with high thrust loads in LO_2 [28]. These bearings used the glass cloth-reinforced PTFE retainer which was chemically treated with HF to improve self-lubrication.

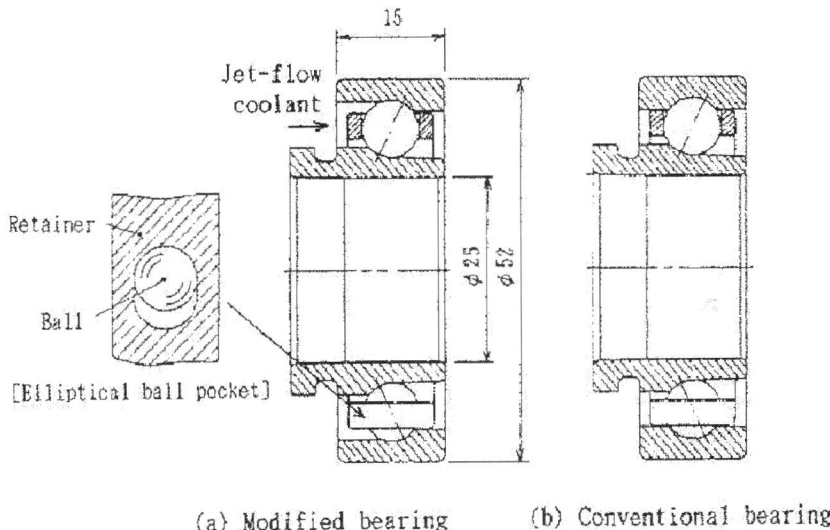

Figure 21. Advanced bearing having single-guided retainer with elliptical ball pocket

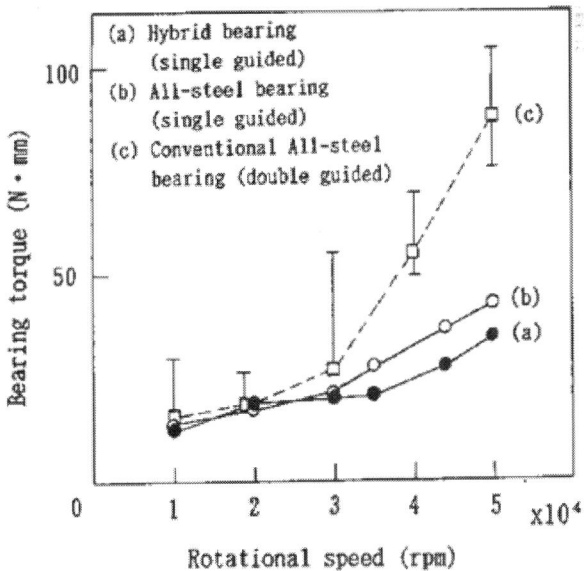

Figure 22. Bearing torque of single-guided bearings and double guided bearing to 50,000 rpm in LH_2

Self–Lubricating Performance And Transfer Film [27,39]
a. *In LH_2*

Figure 22 shows the bearing torque of the single-guided bearings (hybrid ceramic bearing and all-steel bearing) and the conventional double-guided bearing at speeds to 50,000 rpm in LH_2 [39]. It was observed that the bearing torque of the single-guided bearing effectively decreased to about one-half of that of the double-guided bearing. Its result identified that bearing torque induced by high-speed sliding of the outer land guide of the retainer almost accounted for an overall bearing torque generated at high speeds. In addition, the hybrid ceramic bearing showed lower bearing torque than the all-steel bearing at high speeds.

Critical load capacity of the single-guided bearing without a significant rise of the bearing torque and bearing temperature was evaluated. For the single-guided hybrid ceramic bearing tested in LH_2, the critical thrust load was 1,960 N (*Smax* of inner race, 2.7 GPa) at 50,000 rpm and was two times higher than that of the double-guided all-steel bearing. Furthermore, even when bearing torque increased with a rise of bearing temperature, the hybrid ceramic bearing was able to sustain a thrust load of 2,840 N (*Smax*, 3.2 GPa) at 50,000 rpm without seizure in LH_2. High critical load capacity of the single-guided hybrid ceramic bearing was demonstrated [39].

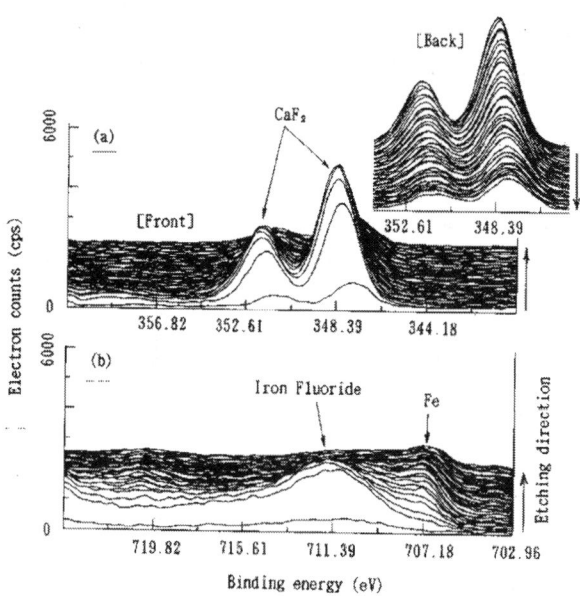

Figure 23. XPS depth analysis of Si_3N_4 ball of hybrid ceramic bearing tested in LH_2

Figure 23 shows the XPS depth analysis of a Si_3N_4 ball taken from the hybrid ceramic bearing tested in LH_2 [27]. Its etching depth was 120 nm (SiO_2 rate). It was found that a considerably thick transfer film consisting of CaF_2/FeF_2 was formed on the ceramic balls. CaF_2 and FeF_2 seemed to be tribo-chemically formed by the reducing power of LH_2. The considerably thick transfer film of CaF_2 and FeF_2 led to exhibit high load capacity. For the all-steel bearing tested in LH_2, a thick CaF_2 film was formed beneath an extremely thin PTFE overlay, but its thickness of CaF_2 transfer film was thinner than that of the hybrid ceramic bearing.

a. *In LO_2*

In LO_2, the hybrid ceramic bearing exhibited poor self-lubricating performance even at a low speed of 10,000 rpm. To the contrary, the all-steel bearing indicated excellent load capacity accompanied by a stable bearing and enabled to sustain a thrust load of 2,650 N (*Smax*, 2.7 GPa) at a speed of 50,000 rpm without seizure in LO_2 [39].

For the hybrid ceramic bearing, an extremely thin, weakly adhesive PTFE film was formed on ceramic balls and resulted in a poor load capacity of the bearing. For the all-steel bearing, the intense formation of a Cr_2O_3 film was beneath an extremely thin PTFE film. It is noted that the tribo-chemical formation of Cr_2O_3 film due to high oxidation power of LO_2 could exhibit high resistance to metal-to-metal adhesion leading to seizure [27].

a. *In LN_2*

The hybrid ceramic bearing exhibited better load capacity than that of the all-steel bearing in LN_2. The hybrid ceramic bearing enabled to sustain a thrust load of 2,700 N (*Smax*, 3.1 GPa) at a speed of 50,000 rpm without seizure. To the contrary, the all-steel bearing showed unstable change of bearing torque and seized at a relatively light-thrust load of 1,470 N (*Smax*, 2.2 GPa) at a speed of 50,000 rpm [39].

For the hybrid ceramic bearing, the thick transfer film consisting of FeF_2/iron oxide formed on the ceramic balls. To the contrary, the seized all-steel bearing was lubricated by only thin PTFE transfer film, without the tribo-chemical formation of $CaF_2/FeF_2/Cr_2O_3$ films because of its inert environment of LN_2. This fact was determined by that the all-steel bearing once tested in LH_2 or LO_2, whose bearing formed the $CaF_2/FeF_2/Cr_2O_3$ films, showed stable change of the bearing torque without seizure even under high thrust loads above 1,470 N in LN_2 [27].

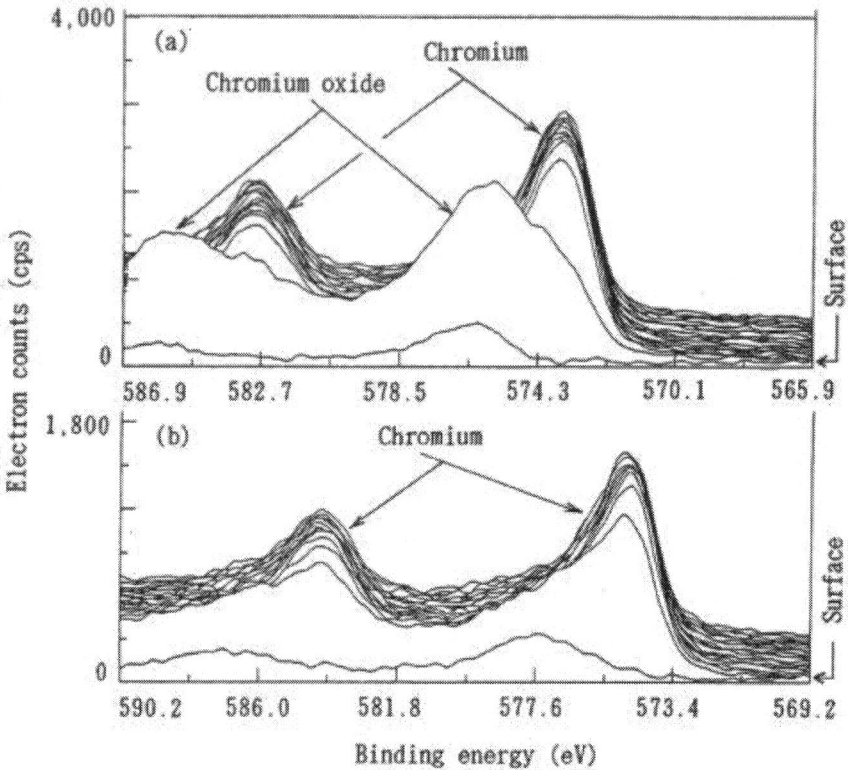

Figure 24. XPS depth analysis of SUS440C ball tested for long run in LO_2 and new ball

Long–Life Bearing [28]
The single-guided all steel bearing was tested for a total operation time to 11.7 hours at a speed of 50,000 rpm with high thrust loads to 2,400 N in LO_2. During long-run test, one-hour operation at a speed of 50,000 rpm was repeated nine times. The test bearing was effectively cooled by the jet-cooling with using nozzles. During the long-run test, the bearing exhibited stable variation of the bearing torque in a range of 93-95 N-mm [28]. The bearing exhibited excellent self-lubrication performance that there was no abnormal change of the bearing torque and bearing temperature.

From the examination of the bearing tested for the long-run test in LO_2, it was observed that sound surface conditions with hardly any wear were determined. The XPS depth analysis of a ball taken from the tested bearing is shown in Fig. 24 [28]. Its etching depth was 30 nm (SiO_2 rate). It is noted that the intense formation of a Cr_2O_3 film was detected and its thickness was thicker than that of the native Cr_2O_3 film on the new ball. Under sufficient cooling conditions in LO_2, the thick Cr_2O_3 film formed by tribo-chemical reaction could provide an extremely high resistance to metal-to-metal adhesion beneath an extremely thin CaF_2 film. To the contrary, under poor cooling conditions in LO_2, the intense formation of oxide film (Fe_2O_3) was mainly produced and led to large mild wear, as discussed in the LO_2 turbopump bearing. Furthermore, the formation of Fe_2O_3 might reduce adhesion of PTFE transfer film, resulting in less lubricant within the bearing. The results indicated that thick formation of a Cr_2O_3 film due to tribo-chemical reaction in LO_2 is important to reduce the bearing wear. Its effect needs sufficient cooling with jet within the bearing components to eliminate the formation of Fe_2O_3 [28].

Fluorine–Passivated Bearing [28]
It is experimentally found that the FeF_2 film formed by a tribo-chemical reaction between the F of PTFE and Fe of 440C steel was facilitated by the high reduction power of LH_2 and enhanced to reduce the bearing wear in LH_2. This may suggest that the FeF_2 film has a good solid-lubricant performance to improve the tribological performance of the bearing. Effect of the coated FeF_2 film on the self-lubrication and durability of the all-steel bearing was evaluated. An FeF_2 film was chemically formed by means of a passivating surface treatment of fluoridation in hot pure F_2 gas. The fluorine-passivated bearings coated with FeF_2 film was tested by long run for 11.7 hours at a speed of 50,000 rpm under high thrust loads in LH_2, LO_2 and LN_2. The fluorine-passivated bearings showed excellent self-lubrication in both LH_2 and LN_2 [28].

In a reduce environment of LH_2, even under poor cooling conditions controlled by reducing of the coolant flow, the fluorine-passivated bearing exhibited superior durability for a total test time to 4.4 hours, as compared with signs of seizure for the untreated bearing. The XPS analysis of the transfer film indicated that the fluorine-passivated bearing was tribo-chemically lubricated by a thick CaF_2 film overlaid on a thick FeF_2/Cr_2O_3 films.

In an inert environment of LN_2, the fluorine-passivated bearing showed excellent self-lubrication and wear conditions for the long-run test up to 11.7 hours at a speed of 50,000 rpm. Stable change of the bearing torque (75-80 N-mm) was shown for the passivated bearing during the long-run test in LN_2 [28]. The bearing test was repeated seven times at a speed of 50,000 rpm and a thrust load of 2,600 N in LN_2. From the examination of the fluorine-passivated bearing tested in LN_2, sound surface conditions with hardly any wear were determined. It was found that a thick CaF_2 film was tribo-chemically formed on thick FeF_2/Cr_2O_3 films of the bearing. On the other hand, the untreated bearing was seized at a low thrust load of 1,470 N due to less tribo-chemical reaction in LN_2, as mentioned before. In such inert environment in LN_2, there was less formation of $CaF_2/FeF_2/Cr_2O_3$ films, so that poor self-lubrication and load capacity of the bearing were shown.

To the contrary, in an oxide environment of LO_2, the fluorine-passivated bearing indicated a higher bearing torque with greater unstable change than that of the untreated bearing [28]. The bearing tests were repeated seven times of the bearing test at a speed of 50,000 rpm and a thrust load of 2,450 N in LO_2. Its total test time was 11.7 hours. During long-run test, high bearing torque continued to vary erratically with the variation in a range of 75-120 N-mm. The fluorine-passivated bearing tested in LO_2 showed somewhat high wear. To the contrary, the untreated bearing demonstrated excellent self-lubrication with hardly any wear during the long-run test as mentioned before. It was clearly showed that the FeF_2 film in LO_2 made a typical reduction in self-lubrication.

Inspection of the fluorine-passivated bearing tested in LO_2 indicated that the initial coated film of FeF_2 was worm away. Its result also indicated that oxide power of LO_2 restricted the tribo-chemical formation of FeF_2 film. Such reduction in self-lubrication possibly resulted from that the coated FeF_2 film restricted the tribo-chemical formation of Cr_2O_3 film in LO_2, resulting in an increase of metal-to-metal adhesion. These results indicated that excellent lubrication depended on the tribo-chemical formation of CaF_2/FeF_2 films in LH_2 or Cr_2O_3 film in LO_2, respectively. In order to obtain high self-lubrication and durability of the bearing, it is noted that tribo-chemical reaction is necessary at the frictional interfaces within the bearing [4].

Ultra–High–Speed Hybrid Ceramic Bearing [8,40]

Based on previous bearing tests at high speeds up to 50,000 rpm, the hybrid ceramic bearing (25-mm bore) was tested at ultra-high-speeds up to 120,000 rpm, and results were compared with the all-steel bearing in LH_2. At a ultra-high speed of 120,000 rpm, the inner-race growth of 34μm due to centrifugal force results in a reduction of the radial clearance within the bearing. Table 4 summarizes comparison of the bearing load and speed conditions for the hybrid ceramic bearing and all-steel bearing at a speed of 120,000 rpm with a thrust load of 980 N [8]. At 120,000 rpm, the initial radial clearance of 77 μm was decreased to 43μm. For the hybrid ceramic bearing, the maximum contact stress *Smax* at the inner race is apt to increase rather than that of the all-steel bearing due to a high elastic modulus. However, the maximum spinning velocity *Vmax* is reduced and resulted in a lower *SVmax* value that leads to a reduction of the bearing temperature and spin wear. The maximum contact stress at the outer race becomes higher due to centrifugal force. For sliding conditions of the retainer, the sliding velocity at the outer land and ball pocket reaches to a high level of 110 m/s and the frictional heat generation of the retainer is to be severe. For the cooling system to remove the bearing heat generation at 120,000 rpm, effective jet cooling with nozzles needs to obtain sufficient coolant flow within the bearing. The nozzles were directed to cool the single outer land-guiding side of the retainer where high frictional heat is generated.

Table 4. Bearing load and speed conditions for hybrid ceramic and all-steel bearings at 120,000 rpm with 980 N (25-mm bore)

Parameters	Hybrid ceramic bearing	All-steel bearing
Bearing		
Rotational speed [rpm]	120,000	
Thrust load [N]	980	
Initial contact angle [deg.]	20	
Initial radial clearance [μm]	77	
Operational radial clearance [μm]	43	
Maximum contact stress at inner/outer races (*Smax*) [GPa]	2.31 / 2.14	2.00 / 2.35
Maximum spinning velocity at inner race (*Vmax*) [m/s]	5.8	7.5
Centrifugal force on ball [N]	454	1,120
Retainer		
Sliding velocity at outer land [m/s]	108	
Sliding velocity at ball pocket [m/s]	116	

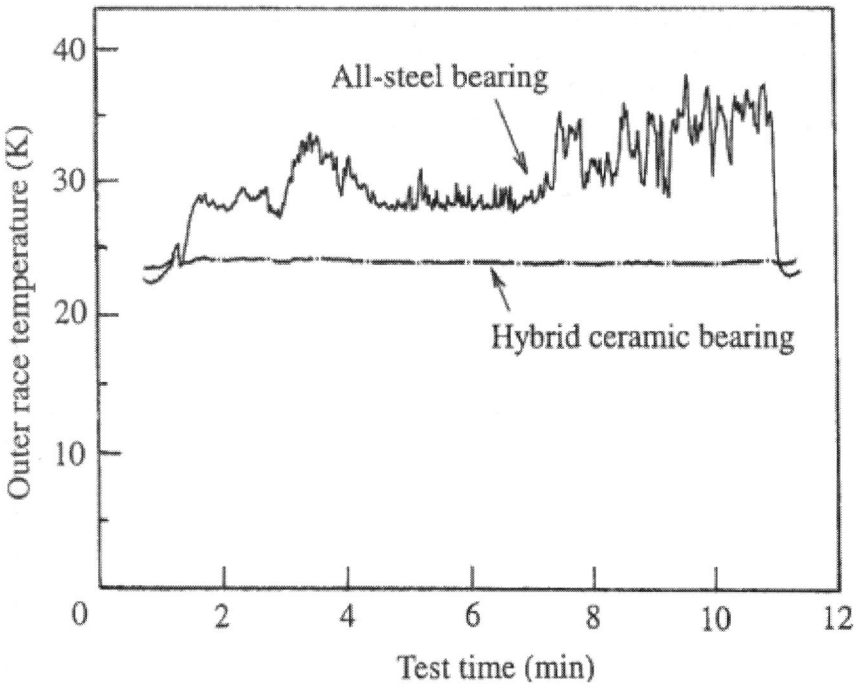

Figure 25. Change of bearing temperature of hybrid ceramic and all-steel bearings at 120,000 rpm with 2,160 N

Figure 25 shows the change of the bearing temperature at a steady speed of 120,000 rpm with a thrust load of 2,160 N [8]. The hybrid ceramic bearing showed excellent performance with a stable condition of the bearing temperature, compared to the seized all-steel bearing showing an irregular change of high bearing temperature. When the thrust load was increased to 3,140 N, the hybrid bearing showed slight damage with a spiky rise of the bearing temperature. It was found that the critical load capacity S_{max} without seizure at a speed of 120,000 rpm was reached to 3.0 GPa (at a thrust load of 2,160 N) for the hybrid ceramic bearing and 2.0 GPa (980 N) for the all-steel bearing, respectively.

The power loss around the bearing was estimated based on the heat absorbed by the cooling flow [8]. Figure 26 shows the power loss of the hybrid ceramic and all-steel bearings as a function of rotational speed up to 120,000 rpm in LH_2 under different cooling conditions at a thrust load of 980 N. It was found that the power loss of the bearing significantly increased above 80,000 rpm with increasing cooling flow rate. At 120,000 rpm, the power loss of the bearing that contained the viscous power loss of

2.2 kW at the shaft side was estimated. The power loss was 6.0 kW for the hybrid ceramic bearing and 6.4 kW for the all-steel bearing, respectively. There was not typical difference of the power loss of the bearing because viscous power loss within the bearing almost accounted for an overall power loss generated at ultra-high speeds. It seems that the power loss around the bearing was mainly induced by viscous drag and churning of the cooling flow passing through the bearing.

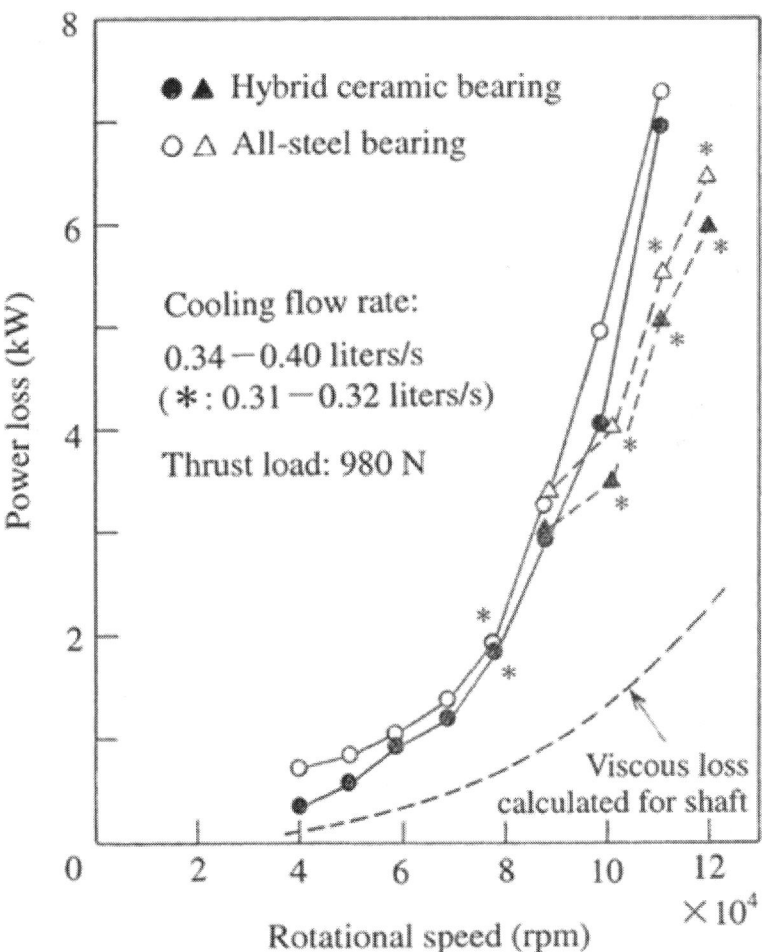

Figure 26. Power loss of hybrid ceramic and all-steel bearings as a function of rotational speed up to 120,000 rpm in LH_2

The components of the hybrid ceramic bearing were in excellent condition with regard to wear at a speed of 120,000 rpm with a thrust load of 3,140 N in LH_2 [40]. On the contrary, the seized all-steel bearing exhibited severe adhesive wear. It was found that the ceramic balls formed superficial micro-cracks on the contact track. Superficial micro-cracks visually extended in a mesh-like pattern on the Si_3N_4 ball tested. It was shown that network of hair crack was propagated along wide-ditch crack. A marked feature of these superficial micro-cracks was that they were very shallow to about 3 μm at maximum and did not extend deeply into the ball. From detailed observation with a scanning electron microscope (SEM), such wide-ditch cracks seemed to be formed by removal of fragments fractured due to contact stress repeated by the rolling balls as shown in Fig. 27. Thus, when the Si_3N_4 balls had lower mechanical strength and fracture toughness, it was clear that wide-ditch cracks were apt to be formed.

An advanced study was conducted to select a tough Si_3N_4 ball capable of restraining crack propagation as well as to evaluate the efficient bearing cooling with nozzles. A Si_3N_4 ball having higher thermal-shock resistance, as well as higher fracture toughness, was found to reduce the propagation of superficial micro-cracks, resulting in a decrease of ball wear. Furthermore, it was observed that the cooling ability of the LH_2 jet-flow aimed at the retainer was superior to that aimed at the inner raceway, further reducing the propagation of thermal micro-cracks on the Si_3N_4 balls. This result also indicated that micro-cracks on the balls were possibly generated at the trace contacting the outer raceway due to a higher centrifugal force under insufficient cooling conditions. Furthermore, under the same cooling rate, the four nozzles achieved a higher cooling ability than the two nozzles with increasing jet speed above 208 m/s. The jet-speed of nozzles reached to the twice of the sliding speed of 108 m/s at the retainer outer-land [40].

In order to prevent the propagation of superficial thermal micro-cracks on the balls, the outer race contact stress was reduced by decreasing the outer race curvature to a limited value of 0.51. Furthermore, sufficient cooling at the outer raceway was gained by a proper clearance of the outer land of the retainer. Decreasing the maximum outer-race stress to 2.0 GPa (thrust load, 1,960 N) in conjunction with sufficient cooling through a narrow outer land clearance could prevent the propagation of superficial micro-cracks even under insufficient cooling conditions [40].

ADVANCED BEARINGS AND SHAFT SEALS

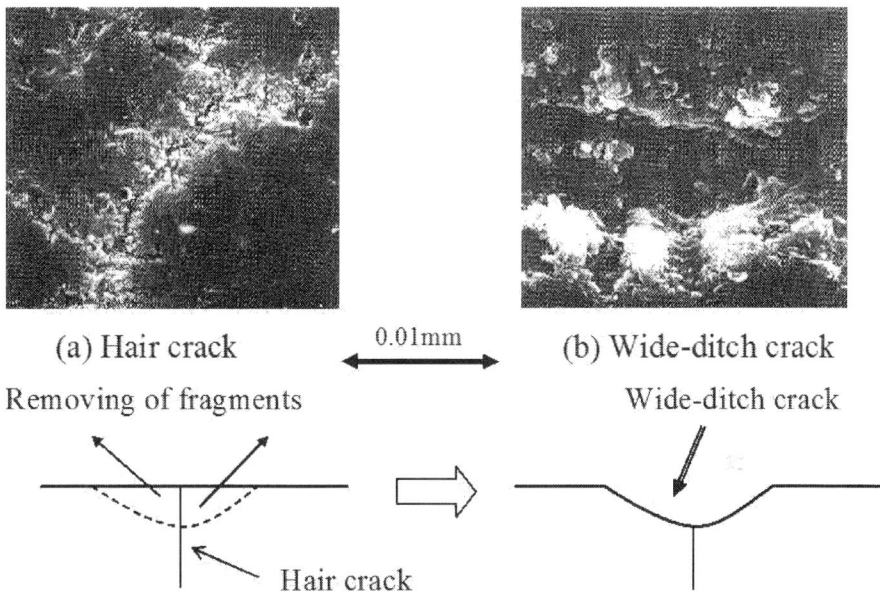

Figure 27. Process model of wide-ditch crack formation on Si_3N_4 ball

Ultra–High–Speed Two–Phase Seal [8]

The floating-ring seal due to noncontact-type is suitable for high-pressure turbopumps; however, conventional seals using carbon seal-rings were weak under high speed and high pressure conditions. Since metal seal-rings have higher mechanical strength and durability, advanced floating-ring seal (with one-seal and two-seal rings) that used Ag-plated metal seal-rings with a seal diameter of 30 mm [8]. This metal seal was studied at ultra-high speeds up to 120,000 rpm in LH_2. Calculated runner growth due to centrifugal force at 120,000 rpm was 29µm, so that the initial seal clearance (gap) was decreased as the rotational speed increased. The test seal had an Ag-plated seal ring made of Inconel 718 that was the same material used for the runner. The runner was coated with a Cr_2O_3 plasma spray, and this coating exhibited excellent friction and wear without adhesion to Ag in LN_2. In order to obtain smooth radial movement of the seal ring, the static seal surface of the housing was coated with a sprayed MoS_2 film.

Figure 28 shows the seal performance of the one-ring seal *vs.* the two-ring seal up to a speed of 100,000 rpm in LH_2 [8]. These seals had a straight bore with a seal gap of 110-120 µm. Figure 29 also shows the phase change models of leakage flow within the seal gap [4]. Seal performance depended on the two-phase flow (gas/liquid phase) of leakage, because the vaporization of leakage was generated by the viscous friction heat and by the seal pressure drop. At lower speeds, the leakage of the one-ring seal was relatively greater than that of the two-ring seal; however, with increasing speed, the leakage of the one-ring seal was drastically decreased and approached the same level of the two-ring seal due to enlargement of the two-phase flow.

For the two-ring seal, the two-phase flow was fully enlarged within the secondary seal ring that was at the downstream of the primary seal ring. Seal leakage was reduced within limits; however, the hydrodynamic force of the liquid phase flow that sustained the seal ring was lost and resulted in seal-ring seizure at a relatively lower speed of 98,700 rpm. Also, shaft vibration for the two-ring seal was likely produced by wobbling of the seal ring under severe rubbing conditions and abruptly increased at speeds of more than 92,000 rpm before resulting in seal-ring seizure at a speed of 98,700 rpm. Furthermore, in the two-ring seal with a seal gap of 70-80 µm, the primary seal-ring seized a speed of 108,600 rpm, because the hydrostatic force decreased due to a low differential pressure.

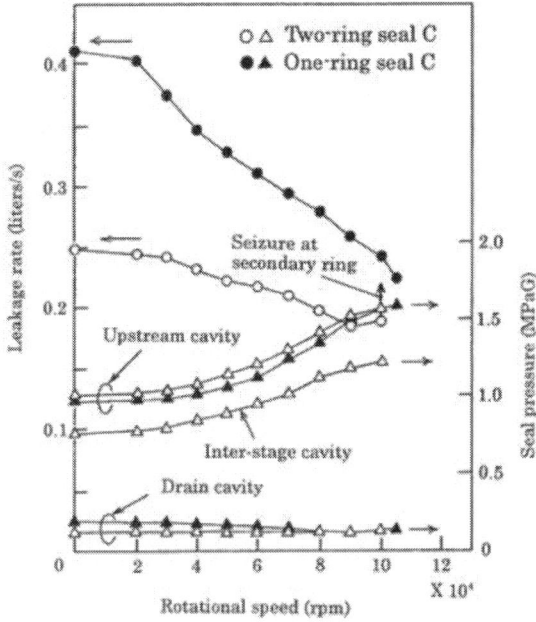

Figure 28. Seal performance of one-ring seal *vs.* two-ring seal up to 100,000 rpm

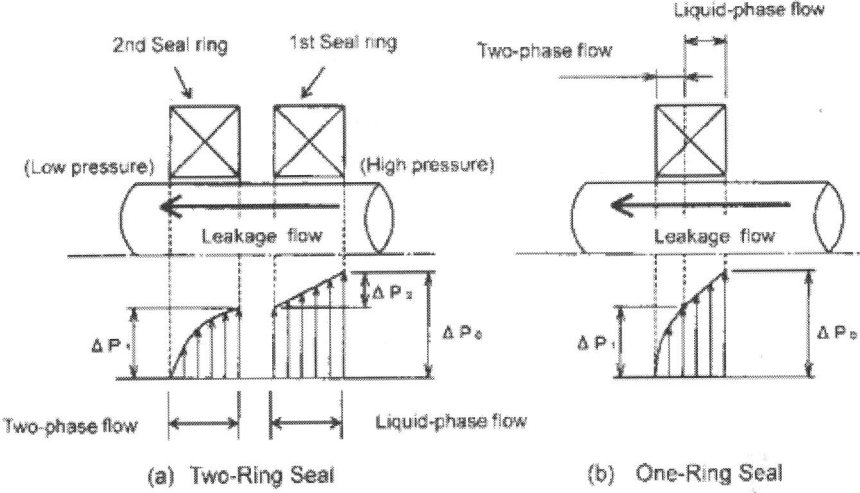

Figure 29. Phase change models of leakage flow within seal gap at ultra-high speed

In contrast, the one-ring seal successfully functioned with no abnormal signs of seizure during tests, because the liquid-phase flow remained within a seal clearance even though the two-phase flow increased. As a result, the hydrodynamic force in the liquid-phase flow as well as the hydrostatic force due to high differential pressure possibly helped to prevent seal-ring seizure. At a steady speed of 120,000 rpm, the one-ring seal exhibited a stable leakage in a range of 0.21-0.24 liters/s that is similar to leakage in the two-ring seal as shown in Fig. 28. Thus, the one-ring seal was superior to the two-ring seal, preventing seal-ring seizure due to an increase of two-phase flow within the sealing clearance.

CONCLUDING REMARKS

For built-up of safe space transport system to achieve high reliability, cryogenic high-speed bearing and shaft seal used in the rocket turbopumps are reviewed historically. These tribo-components have specific lubrication, materials and design requirements in pumping cryogenic liquid propellants in rocket engines. Nowadays, as earth scale issues of energy conservation and environment preservation, a breakaway from the conventional fossil-fuel society becomes a big problem. Clean hydrogen energy is attractive due to its energy efficiency and its smaller impact on the environment, and it is expected to be a key technology in the 21st century. It is proposed that, to build hydrogen infrastructure for

LH_2 storage and distribution, development of an industrial tribo-system with long durability and high reliability is essential and advances by supporting of cryogenic tribology studied for LH_2 rocket system.

ACKNOWLEDGEMENTS

This paper is based on previous cryogenic tribology studies carried out by Japan Aerospace Exploration Agency (JAXA) at Kakuda Space Center. These studies were also supported by IHI Corporation for turbopumps, by NTN corporation for bearings and by Eagle Industry Co., LTD. for shaft seals, respectively. The author is indebted to researchers engaged for their valuable support, to organizations for their enthusiastic cooperation. At last, the author has to thank late Prof. Miyakawa, Y. of Hhosei University, as a pioneer in space tribology in Japan, for his guidance to cryogenic tribology with profound appreciation.

REFERENCES

1. Dieter K H & David H H. Modern Engineering for Design of Liquid-Propellant Rocket Engines, *Progress in Astronautics and Aeronautics, Vol. 147*, AIAA, (1992), 155-218.
2. Liquid Rocket Engine Turbopump Bearing, NASA SP-8048, 1971.
3. Liquid Rocket Engine Turbopump Rotating- Shaft Seals, NASA SP-8121, 1978.
4. Nosaka, M. Cryogenic Tribology of High-Speed Bearings and Shaft Seals in Liquid Hydrogen, *Tribology Online*, 6, 2, (2011), 133-141.
5. Nosaka M, Takada S & Yoshida M. Research and Development of Cryogenic Tribology of Turbopumps for Rocket Engines, *J. of Aeronautical and Space Science Japan*, 58, 681, (2010), 303-313, in Japanese.
6. Hale J R & Klatt F T. SSME Improvement for Routine Shuttle Operations, AIAA-85-1163, (1985).
7. Gibson H. Lubriction of Space Shuttle Main Engine Turbopump Bearings, *Lubr. Eng.* 57, 8, (2001), 10-12.
8. Nosaka M, Takada S, Kikuchi M, Sudo T & Yoshida M. Ultra-High-Speed Performance of Ball Bearings and Annular Seals in Liquid Hydrogen at Up to 3 Million DN (120,000 rpm), *Trib. Trans.*, 47, (2004), 43-53.

REFERENCES

9. Ohta T, Kimoto K, Kawai T, Motomura T, Russ M & Paulus T. Design, Fabrication and Test of the RL60 Fuel Turbopump, AIAA-2003-5073, (2003).
10. Collongeat L, Edeline E, Frocot M & Dehouve J. Development status of high DN LH2 bearings in Snecma, AIAA-2005-3950, (2005).
11. Rachuk V & Titkov N. The First Russian LOX-LH2 Expander Cycle LRV: RD0146, AIAA-2006-4904, (2006).
12. Nosaka M, Oike M, Kamijo K, Kikuchi M & Katsuta H. Experimental Study on Lubricating Performance of Self-Lubricating Ball Bearings for Liquid Hydrogen Turbopump, *Lubr. Eng.*, 44, 1, (1988), 30-44.
13. Nosaka M, Oike M, Kikuchi M, Kamijo K & Tajiri M. Tribo-Characteristics of Self-Lubricating Ball Bearings for the LE-7 Liquid Hydrogen Rocket-Turbopump, *Trib. Trans.*, 36, 3, (1993), 432-442.
14. Nosaka M, Oike M, Kikuchi M, Nagao R & Mayumi T. Evaluation of Durability for Cryogenic High-Speed Ball Bearings for LE-7 Rocket Turbopumps, *Lubr. Eng.*, 52, 3, (1996), 221-233.
15. Nosaka M, Oike M, Kikuchi M, Kamijo K & Tajiri M. Self-Lubricating Performance and Durability of Ball Bearings for the LE-7 Liquid Oxygen Rocket-Turbopump, *Lubr. Eng.*, 49, 9, (1993), 677-688.
16. Nosaka M, Oike M & Kikuchi M. Tribology at Low and High Temperatures, Lubrication in Rocket-Turbopumps, *J. of Japan Society of Lubrication Engineers*, 33, 2, (1988), 90-96, in Japanese.
17. Nosaka M. Self-Lubricating Performance of High-Speed Ball Bearing for Liquid Hydrogen (1), Design Problems, *J. of Japan Society of Lubrication Engineers*, 32, 10, (1987), 689-695, in Japanese.
18. Nosaka M. Self-Lubricating Performance of High-Speed Ball Bearing for Liquid Hydrogen (2), Self-Lubricating Performance Improvements. *J. of Japan Society of Lubrication Engineers*, 32, 12, (1987), 833-838, in Japanese.
19. Winn L W, Eusepi M W & Smalley A J. Small, High-Speed Bearing Technology for Cryogenic Turbo-Pumps, NASA CR-134615, 1974.
20. Edmond E B & William J A. Advanced Bearing Technology, NASA SP-38, 1965.
21. Nosaka M & Oike M. Rotating-Shaft Seals in Rocket-Turbopumps, *J. of Japanese Society of Tribologists*, 35, 4, (1990), 233-238, in Japanese.
22. Oike M, Nosaka M, Watanabe Y, Kikuchi M & Kamijo K. Experimental Study on High-Pressure Gas Seals for a Liquid Oxygen Turbopump, *STLE Trans.*, 31, 1, (1988), 91-97.
23. Nosaka M, Oike M & Kikuchi M. Cryogenic Tribology of Turbopumps for Rockets, *Cryogenic Engineering*, 31, 10, (1996), 500-511, in Japanese.
24. Nosaka M. Tribological Burn-Out of Wear, *J. of Japanese Society of Tribologists*, 36, 9, (1991), 689-691, in Japanese.

25. Nosaka M. Tribology in Low Temperature Environment, *J. of Japanese Society of Tribologists*, 52, 11, (2007), 759-764, in Japanese.
26. Nosaka M, Takada S, Yoshida M, Kikuchi M, Sudo T & Nakamura S. Effect of Tilted Misalignment of Tribo-Characteristics of High-Speed Ball Bearings in Liquid Hydrogen, *Tribology Online*, 5, 2, (2010), 71-79.
27. Nosaka M, Kikuchi M, Oike M & Kawai N. Tribo-Characteristics of Cryogenic Hybrid Ceramic Ball Bearings for Rocket Turbopumps: Bearing Wear and Transfer Film, *Trib. Trans.*, 42, 1, (1999), 106-115.
28. Nosaka M, Kikuchi M, Kawai N & Kikuyama H. Effect of Iron Fluoride Layer on Durability of Cryogenic High-Speed Ball Bearings for Rocket Turbopumps, *Trib. Trans.*, 43, 2, (2000), 163-174.
29. Suzuki M, Nosaka M, Kamijo K & Kikuchi M. Research and Development of a Rotating- Shaft Seals for a Liquid Hydrogen Turbopump, *Lubr. Eng.*, 42, 3, (1986), 162-169.
30. Nosaka M, Miyakawa Y, Kamijo K, Suzuki M & Kikuchi M. Study on Sealing Characteristics of High Speed, Contacting Mechanical Seals for Liquid Hydrogen (Part 1), Development of Mechanical Seal for Liquid Hydrogen Turbopump, *J. of Japan Society of Lubrication Engineers*, 29, 1, (1984), 35-42, in Japanese.
31. Nosaka M, Kamijo K, Suzuki M, Kikuchi M & Miyakawa Y. Study on Sealing Characteristics of High Speed, Contacting Mechanical Seals for Liquid Hydrogen (Part 2), Starting Torque and Static Sealing Performance, *J. of Japan Society of Lubrication Engineers*, 29, 1, (1984), 43-49, in Japanese.
32. Nosaka M, Kamijo K, Suzuki M, Kikuchi M & Miyakawa Y. Study on Sealing Characteristics of High Speed, Contacting Mechanical Seals for Liquid Hydrogen (Part 3), Friction Power Loss and Dynamic Sealing Performance, *J. of Japan Society of Lubrication Engineers*, 29, 2, (1984), 113-120, in Japanese.
33. Nosaka M, Kamijo K, Suzuki M, Kikuchi M & Miyakawa Y. Study on Sealing Characteristics of High Speed, Contacting Mechanical Seals for Liquid Hydrogen (Part 4), Characteristics of Running Process and Wear of Rubbing Seal Faces, *J. of Japan Society of Lubrication Engineers*, 29, 2, (1984), 121-128, in Japanese.
34. Nosaka M, Kamijo K, Suzuki M, Kikuchi M & Miyakawa Y. Study on Sealing Characteristics of High Speed, Contacting Mechanical Seals for Liquid Hydrogen (Part 5), The Formation of Thermal Crack and Wear in Chromium Plate on Rotating Ring, *J. of Japan Society of Lubrication Engineers*, 29, 3, (1984), 187-194, in Japanese.
35. Oike M, Nosaka M, Kikuchi M & Watanabe Y. Performance of A Shaft Seal System for The LE-7 Rocket Engine Oxidizer Turbopump, *Proc. of The 18th Inter. Symposium on Space Tech. and Sci., Kagoshima*, (1992), 143-154.

36. Oike M, Nosaka M, Kikuchi M & Hasegawa S. Two-Phase Flow in Floating-Ring Seals for Cryogenic Turbopumps, *Tribo. Trans.*, 42, 2, (1999), 273-281.
37. Oike M, Nosaka M, Kikuchi M & Watanabe Y. Performance of a Segmented Circumferential Seal for a Liquid Oxygen Turbopump (Part 1), Sealing Performance, *J. of Japanese Society of Tribologists*, 37, 4, (1992), 339-346, in Japanese.
38. Oike M, Nosaka M, Kikuchi M & Watanabe Y. Performance of a Segmented Circumferential Seal for a Liquid Oxygen Turbopump (Part 2), Durability, *J. of Japanese Society of Tribologists*, 37, 5, (1992), 389-396, in Japanese.
39. Nosaka M, Kikuchi M, Oike M & Kawai N. Tribo-Characteristics of Cryogenic Hybrid Ceramic Ball Bearings for Rocket Turbopumps: Self-Lubricating Performance, *Trib. Trans.*, 40, 1, (1997), 21-30.
40. Nosaka M, Takada S, Yoshida M, Kikuchi M, Sudo T & Nakamura S. Improvement of Durability of Hybrid Ceramic Ball Bearings in Liquid Hydrogen at 3 Million DN (120,000 rpm), *Tribology Online*, 5, 1, (2010), 60-70.

CITATION

Masataka Nosaka and Takahisa Kato (2013). Cryogenic Tribology in High-Speed Bearings and Shaft Seals of Rocket Turbopumps, Tribology - Fundamentals and Advancements, Dr. Jürgen Gegner (Ed.), ISBN: 978-953-51-1135-1, InTech, DOI: 10.5772/55733.

CHAPTER 7

Surface Modification by Friction Based Processes

R. M. Miranda[1], J. Gandra[2] and P. Vilaça[2]

[1] Mechanical and Industrial Engineering Department, Sciences and Technology Faculty, Nova University of Lisbon, Caparica, Portugal
[2] Mechanical Engineering Department, Lisbon Technical University, Av. Rovisco Pais, Lisboa, Portugal

INTRODUCTION

The increasing need to modify the surface's properties of full components, or in selected areas, in order to meet with design and functional requirements, has pushed the development of surface engineering which is largely recognised as a very important field for materials and mechanical engineers.

Surface engineering includes a wide range of processes, tailoring chemical and structural properties in a thin surface layer of the substrate, by modifying the existing surface to a depth of 0.001 to 1.0 mm such as: ion implantation, sputtering to weld hardfacings and other cladding processes, producing typically 1 - 20 mm thick coatings, usually for wear and corrosion resistance and repairing damaged parts. Other deposition processes, such as laser alloying or cladding, thermal spraying, cold spraying, liquid deposition methods, anodising, chemical vapour deposition (CVD), and physical vapour deposition (PVD), are also extensively used in surface engineering. Hardening by melting and rapid solidification and surface mechanical deformation allow to change the properties without modifying its composition [1].

Friction based processes comprise two manufacturing technologies and these are: Friction Surfacing (FS) and Friction Stir Processing (FSP). The former was developed in the 40´s [2] and was abandoned, at that time, due to the increasing developments observed in competing technologies as thermal spraying, laser and plasma. Specially laser surface technology has largely developed in the following years in hardening, alloying and cladding applications and is now well established in industry. However, FS as a solid state processing technology, was brought back for thermal sensitive materials due to its possibility to transfer material from a consumable rod onto a substrate producing a coating with a good bonding and limited dilution.

The patented concept of Friction Stir Welding in the 90´s [5] opened a new field for joining metals, specially light alloys and friction stir processing emerged around this concept.

FSP uses the same basic principles as friction stir welding for superficial or in-volume processing of metallic materials. Applications are found in localized modification and microstructure control in thin surface layers of processed metallic components for specific property enhancement. It has proven to be an effective treatment to achieve major microstructural refinement, densification and homogenisation of the processed zone, as well as, to eliminate defects from casting and forging [6-8]. Processed surfaces have enhanced mechanical properties, such as hardness, tensile strength, fatigue, corrosion and wear resistance. A uniform equiaxial fine grain structure is obtained improving superplastic behaviour. FSP has also been successfully investigated for metal matrix composite manufacturing (MMCs) and functional graded materials (FGMs) opening new possibilities to chemically modify the surfaces [9].

However, FSP has some disadvantages, the major of which is tool degradation and cost, which limits its wider use to high added value applications. Therefore, friction surfacing (FS) emerged again.

This chapter will focus on the mechanisms involved in both FSP and FS and their operating parameters, highlighting existing and envisaged applications in surface engineering, based on the knowledge acquired from ongoing research at the author's institutions.

FRICTION STIR PROCESSING

Fundamentals

Friction Stir Processing (FSP) is based on the same principles as friction stir welding (FSW) and represents an important breakthrough in the field of solid state materials processing.

FSP is used for localized modification and microstructural control of surface layers of processed metallic components for specific property enhancement [6]. It is an effective technology for microstructure refinement, densification and homogenisation, as well as for defect removal of cast and forged components as surface cracks and pores. Processed surfaces have shown an improvement of mechanical properties, such as hardness and tensile strength, better fatigue, corrosion and wear resistance. On the other hand, fine microstructures with equiaxed recrystallized grains improve superplastic behaviour of materials processing and this was verified for aluminium alloys [7]. More recently the introduction of powders preplaced on the surface or in machined grooves allowed the modification of the surfaces, producing coatings with characteristics different from the bulk material, or even functionally graded materials to be discussed later in this chapter. The process has still limited industrial applications but is promising due to its low energy consumption and the wide variety of coating / substrate material combinations allowed by the solid state process.

A non-consumable rotating tool consisting of a pin and a shoulder plunges into the workpiece surface. The tool rotation plastically deforms the adjacent material and generates frictional heat both internally, at an atomic level, and between the material surface and the shoulder. Localized heat is produced by dissipation of the internal deformation energy and interfacial friction between the rotating tool and the workpiece. The local temperature of the substrate rises to the range where it has a viscoplastic behaviour beneficial for thermo-mechanical processing. When the proper thermo-mechanical conditions, necessary for material consolidation are achieved, the tool is displaced in a translation movement. As the rotating tool travels along the workpiece, the substrate material flows, confined by the rigid tool and the adjacent cold material, in a closed matrix like forging manufacturing process. The material under the tool is stirred and forged by the pressure exerted by the axial force applied during processing as depicted in Figure 1.

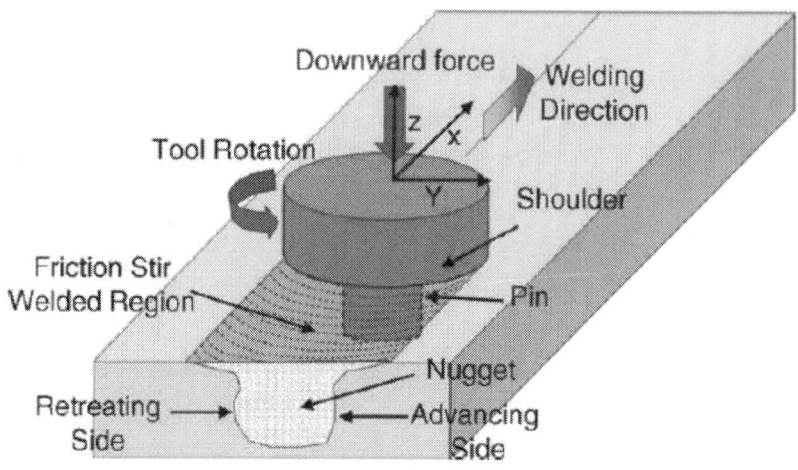

Figure 1. Schematics of friction stir processing.

The material structure is refined by a dynamic recrystallization process triggered by the severe plastic deformation and the localised generated heat. Homogenization of the structure is also observed along with a defect free modified layer of micrometric or nanometric grain structure.

FSP is considered an environmentally friend technology due to its energy efficiency and absence of gases or fumes produced. Table 1 summarizes the major benefits of FSP considering technical, metallurgical, energy and environment aspects.

Table 1. Major benefits of friction stir processing

Technical	Processed depth controlled by the pin length
	One-step technique
	No surface cleaning required
	Good dimensional stability since it is performed under solid state
	Good repeatability
	Facility of automation
Metallurgical	Solid state process
	Minimal distortion of parts
	No chemical effects
	Grain refining and homogenization
	Excellent metallurgical properties
	No cracking
	Possibility to treat thermal sensitive materials
Energy	Low energy consumption since heat is generated by friction and plastic deformation
	Energy efficiency competing with fusion based processes as laser
Environmental	No fumes produced
	Reduced noise
	No solvents required for surface degreasing and cleaning

Analysing the cross section of a friction stir processed surface, three distinct zones can be identified and these are: the nugget or the stirred zone (SZ), the thermomechanically affected zone (TMAZ) and the heat affected zone (HAZ) as shown in Fig. 2.

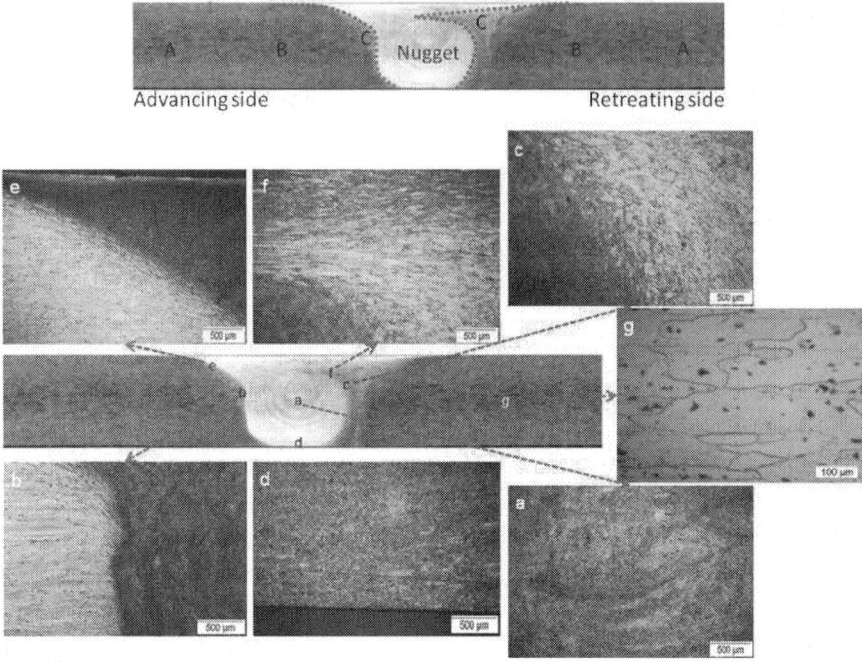

Figure 2. A typical macrograph showing various microstructural zones in FSW of AA2024-T351

The nugget, just below the pin and confined by the shoulder width, is the area of interaction where severe plastic deformation occurs. The raise in local temperature due to internal friction and the generated friction between the shoulder and the surface along with the high strain, promotes a dynamically recrystallized zone, resulting in the generation of fine homogeneous equiaxial grains in the stirred zone and precipitate dissolution. Though this is a solid state process, the maximum temperature can be of about 80% of the fusion temperature. Ultrafine-grained microstructures with an average grain size of 100-300 nm in a Mg-Al-Zn alloy were observed in a single pass under cooling [8]. These micro and nano structures are responsible for increases in hardness and wear behaviour reported by several researchers studying different types of alloys under different processing conditions.

The thermo-mechanically affected zone (TMAZ) is immediately adjacent to the previous and, in this zone, the deformation and generated heat are insufficient to generate new grain formation, thus, deformed elongated grains are observed with second phases dispersed in the grain boundaries.

Though new grain nucleation may be observed, microstructure remains elongated and deformed. The hardness is higher than in the heat-affected zone due to the high dislocations density and sub-boundaries caused by plastic deformation.

In the heat affected zone (HAZ) no plastic deformation is experienced but heat dissipated from the stirred zone into the bulk material can induce phase transformations, depending on the alloys being processed, as precipitate coarsening, localized aging or annealing phenomena.

The non symmetrical character of the process is also evident (Fig. 2). The advancing side is usually referred to the one where the rotating and travel movements have the same direction, while in the retreating side these have opposite directions. On the advancing side, the recrystallized zone is extended and the nugget presents a sharp appearance. The relative velocity between the tool and the base material is higher due to the combination of tool rotation and translation movement. As such, plastic deformation is more intense, thus, the increase in the degree of deformation during FSP, results in a reduction of recrystallized grain size, extending the fine-grain nugget region to the advancing side. Hardness can be higher than in the thermo-mechanically affected zone, but typically lower than in the base material, whenever it is a heat treatable alloy hardenable by aging. The Hall Petch equation establishes a relation between grain size and yield strength and states that these vary in opposite senses [10]. So, in the nugget yield strength is seen to be much higher than in the base material and this is a major result from this process.

Processing Parameters

Operating or processing parameters determine the amount of plastic deformation, generated heat and material flow around the non-consumable tool.

The tool geometry is of major relevance as far as material flow is concerned. Two main elements constitute the tool and these are the pin and the shoulder. Geometrical features such as pin height and shape, shoulder surface pattern and diameter, have a major influence on material flow, heat generation and material transport volume, determining the final microstructure and properties of a processed surface. Several tools have been designed and patented for both FSW and FSP.

The pin (Fig. 3) can be cylindrical or conical, flat faced, threaded or fluted to increase the interface between the probe and the plasticized material, thus intensifying plastic deformation, heat generation and material mixing. The pin length determines the depth of the processed layer. However, since FSP usually aims to produce a thin fine-grained layer across a larger surface area, pinless tools with larger shoulder diameters can also be used.

Figure 3. Example of tool geometries

Shoulder profiles aim to improve friction with the material surface generating the most part of frictional heat involved in the process. Shoulders can be concave, flat or convex, with grooves, ridges, scrolls or concentric circles as depicted in Figure 4.

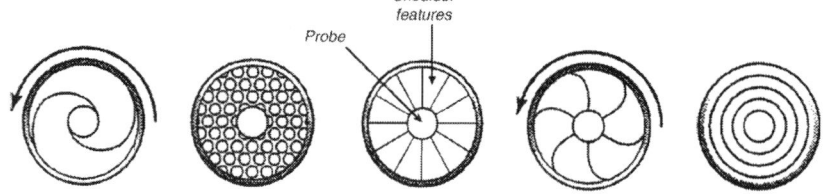

Figure 4. Examples of shoulder geometries

The major processing parameters are the tool rotation and traverse speeds, the axial force and the tilt angle:

- Tool rotation and traverse speeds

These parameters, individually or in combination, affect the plastic deformation imposed onto the material and, thus, the generated heat. An empirically accepted concept divides processing into two main classifications: "cold" and "hot". Cold processing is the one where the ratio between rotating and traverse speed is below 3 rpm/mm and hot processing when this ratio is above 6. Though there is no scientific basis for this border line, it is, however, noticeable that increasing this ratio, the SZ is larger and a very fine structure is observed, while under "cold" conditions the SZ is not well defined since the heat generated is insufficient to promote grain recrystallization. So, increasing the tool rotation speed, plastic deformation is more intense and so is generated heat enabling more material mixing. Therefore, it is possible to achieve a smaller grain size of equiaxial homogeneous grains with precipitate dissolution.

Transverse speed mostly affects the exposure time to frictional heat and material viscosity. Low traverse speeds result in larger exposure times at higher process temperatures.

- Tool axial force

This parameter affects friction between the shoulder and the substrate surface generating and promoting material consolidation. High axial force causes excessive heat and forging pressure, obtaining grain growth and coarsening, while low axial forces lead to poor material consolidation, due to insufficient forging pressure and friction heating. Excessive force may also result in shear lips or flashes with excessive height of the beads on both the advancing and retreating sides, causing metal thinning at the processed area and poor yield and tensile properties. So, surface finishing is much controlled by the axial or forging force.

- Tilt angle

The tilt angle is the angle between the tool axis and the workpiece surface. The setting of a suitable tilting towards the traverse direction assures that the tool moves the material more efficiently from the front to the back of the pin and improves surface finishing.

The effect of the different process parameters has been widely documented by several authors and they are all unanimous that plastic deformation and consequent heat generation are essential to establish the viscoplastic conditions necessary for the material flow and to achieve good consolidation. Thus, a tilt angle of 2-4° is usually used in practice.

Insufficient heating, caused by poor stirring (low tool rotational rates), a high transverse speed or insufficient axial force, results in improper material consolidation with consequent low strength and ductility. Raising heat will cause grain size to decrease to a nanometric scale improving material properties. However, a very significant increase in tool rotation rate, axial force or a very low transverse speed may result in high non desired temperature, slow cooling rate or excessive release of stirred material with property degradation.

Multiple Passes

In order to process large areas in full extent, multiple-passes are used. These can be run separately or overlapped. An overlap ratio (OR) was defined to characterize the overlap between passes and defined by equation 1 [7].

$$OR = 1 - \left[\frac{l}{d_{pin}}\right]$$

(1)

Where *l* is distance between centres of each pass and *dpin* is the maximum diameter of the pin. From this equation, fully overlapped passes have an OR=1 and OR decreases, when increasing the distance between passes. For an OR<0 no overlap of the nuggets exists.

There are two types of material modification by Friction Stir Processing, the in-volume FSP (VFSP) consisting on the modification of the full thickness of the processed materials and the surface FSP (SFSP) which consists in the surface modification up to depth of about 2 mm.

Figure 5 depicts the effect of OR in two Al alloys, a heat treatable (AA7022-T6) and a non heat treatable one (AA5083-O) with different number of passes and overlap ratios.

FRICTION STIR PROCESSING

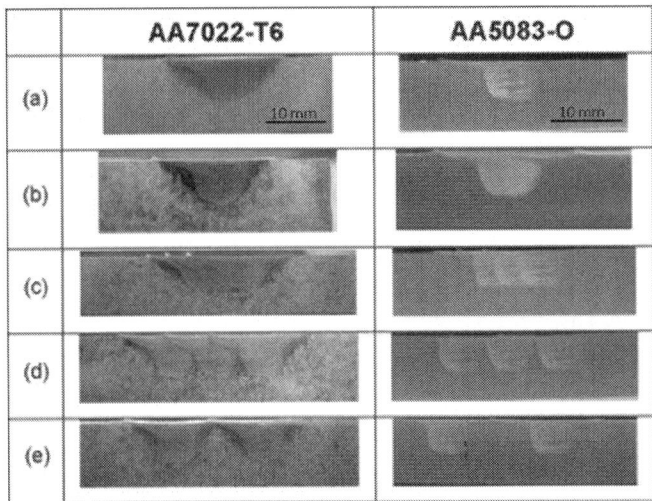

Figure 5. Cross sections of the samples processes with different treatments a) one pass with OR=1; b) four passes with OR=1; c) three passes with OR=1/2; d) three passes with OR=0 and e) two passes with OR=-1 [7]

In this study [7] the authors showed that AA5083-O alloy needed at least three passes in the same location to produce a homogeneous processed area, while the AA7022-T6 alloy only needed one pass, since this is a heat treatable alloy. Grain size reduced from 160 μm (AA7022-T6) and 106 μm (AA5083-O) to an average grain size of about 7.1 and 5.9 μm, respectively. The highest hardness value was located in the nugget due to a significantly decrease in the grain size. This results that in AA7022-T6 alloy the hardness is lower in the nugget than in the base material because it is a heat treatable aluminium alloy and in the AA5083-O alloys the hardness in the nugget is higher than in the base material which is a typical behaviour of non-heat treatable alloys. A significant increase in the formability of the materials was observed due to the increase of the materials ductility resulting from the refinement of the grain size, increasing the maximum bending angle in four times for the SFSP treatment and twelve times for the VFSP treatment in the AA7022-T6 samples. In AA5083-O samples an increase in the maximum bending angle around 1.5 times for the SFSP treatment and about 2.5 times for VFSP treatment was observed.

The overlapping direction in multipass Friction Stir Processing (FSP) was also seen to have a major influence on the surface geometrical features [11]. Structural and mechanical differences were observed in a AA5083-

H111 alloy when overlapping by the advancing side (AS) direction or by the retreating side (RS) one. Overlapping by the retreating side was found to generate smoother surfaces, while overlapping by the advancing side led to more uniform thickness layer (Fig. 6). This result is quite relevant from a practical point of view since when the aim of processing large areas in multiple passes procedure is to increase the depth of the processed zone, overlapping of successive passes should be performed by the advancing side of the previous pass. If surface finishing is to be maximised to prevent finishing operations, overlapping on the previous pass in the retreating side produces very low rough surfaces.

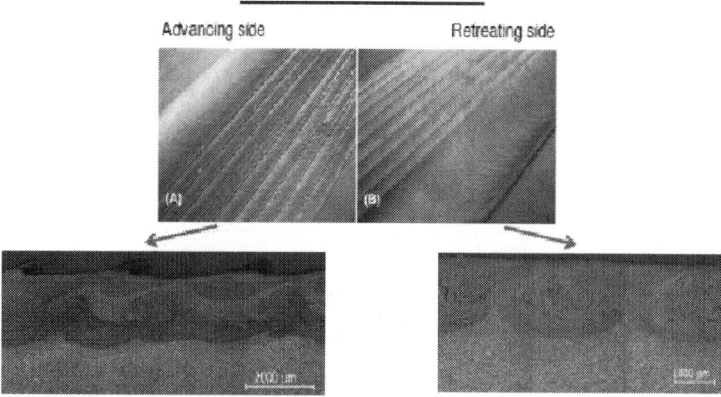

Figure 6. Macro and micrographs of cross sections in friction stir processed surfaces when overlapping by the advancing and by the retreating sides [11]

Hardness within the processed layer increased by 8.5 % and was seen to be approximately constant between passes. The mechanical resistance and toughness under bending were improved by 18 % and 19 %, respectively.

Bending test curves are presented in Figure 7. The processed surfaces were tested under tensile and compression loads. Different behaviours were observed for each bending specimen. Surface modification by multi-pass FSP resulted in an increase of the maximum load supported for all samples and up to a maximum of 18 % for the compression solicitation of the sample produced when overlapping by the RS (Fig. 8). FSP produced a thin layer of a fine equiaxial recrystallized grain structure and homogeneous precipitation dispersion, enhancing material strength.

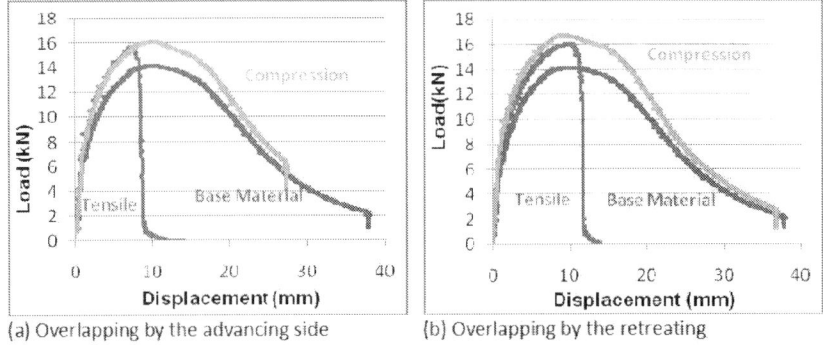

Figure 7. Load vs. displacement plot of the bending tests of the FSP samples produced when overlapping by (a) AS and (b) RS [11].

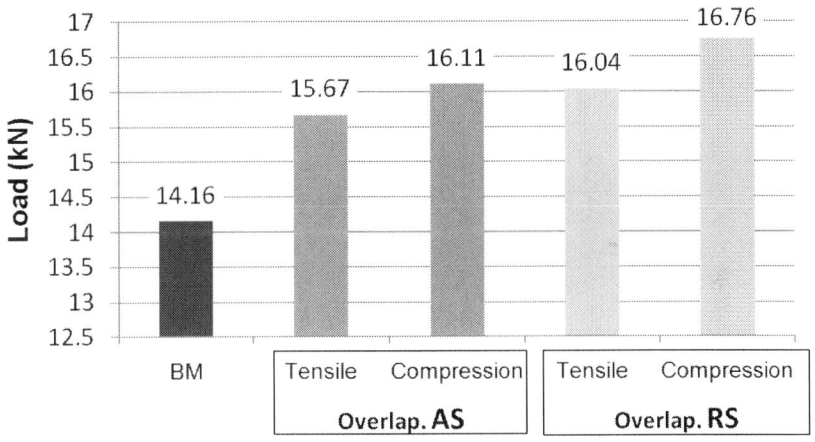

Figure 8. Maximum load attained by FSP samples under different test conditions relatively to the base material [11].

Applications and Performance

Friction stir processing can be used to locally refine microstructures and eliminate casting defects in selected locations, where property improvements can enhance component performance and service lifetime. For instances, aluminium castings contain porosities, segregated phases and inhomogeneous microstructures which contribute to property degradation. Microstructural casting defects, such as: coarse precipitates and porosities increase the possibility of rupture due to the intragranular nucleation of micro-cracks during material deformation. Precipitates are less capable of plastic deformation than the matrix, so cavity nucleation is very frequent, whether caused by a disconnection from the matrix or the rupture of precipitates.

Friction stir processing allows the breakage of large precipitates and their dispersion in a homogeneous matrix, increasing the material capability to withstand deformation, since it results in a higher level of crack closure. Additionally, mechanical properties such as ductility, fatigue strength and formability, are improved.

On the other hand, a large number of small precipitates increases the material resistance to deformation and hence its strength, as they act as barriers or anchorage points to dislocations movements. A uniform equiaxial fine grain structure is also essential to enhance material superplastic behaviour. Friction stir processing generates fine microstructure and equiaxed recrystallized grains which leads either to an increase in strain rate or a decrease in the temperature at which superplasticity is achieved.

In the FSP of an aluminium cast alloy ADC12, Nakata et al. [12] applied multiple-passes to increase tensile strength to about 1.7 times that of the base material. The hardness profile of processed layer was uniform and about 20 HV higher than that of the cast material. The observed increase in tensile strength was attributed to the elimination of the casting defects such as porosities, an homogeneous redistribution of fine Si particles and a significant grain refinement to 2–3 μm. Santella et al. [13] investigated the use of friction stir processing to homogenise hardness distributions in A319 and A356 cast aluminium alloys. Hardness and tensile strength were increased relatively to the cast base material.

Similar results were also reported in the friction stir processing of magnesium based alloys. A.H. Feng and Z.Y. Ma et al. [14] combined FSP with subsequent aging to enhance mechanical properties of Mg-Al-Zn castings

Chang et al. [8] obtained a significant improvement of mechanical properties as the mean hardness measured at the ultrafine-grained zone reached approximately 120HV (twice the base material hardness).

Several investigations have been conducted to study the enhancement of superplasticity behaviour in friction stir processed alloys. In the FSP of Al-8.9Zn-2.6Mg-0.09Sc, Charit and Mishra et al. [15] reported a maximum superplasticity of 1165% at a strain rate of 3×10^{-2} s^{-1} and 310 °C with a grain size of 0.68 μm. More recently, F.C. Liu [16] reported a fine-grain microstructure of 2.6 μm sized grains by applying FSP to extruded samples of an Al-Mg-Sc alloy, achieving a maximum elongation of 2150% at a high strain rate of 1×10^{-1} and a temperature of 450 °C.

A ultrafine-grained FSP Al-Mg-Sc alloy was also reported [17] with a grain size of 0.7 μm exhibited high strain rate superplasticity, for a low temperature range of 200 to 300 °C with a single pass. For a strain rate of 3×10^{-2} s^{-1} at a temperature of 300 °C, a maximum ductility of 620% was achieved. However, for a temperature of 350 °C, abnormal grain growth was observed, as grain size increased and the samples no longer presented superplasticity, thus confirming that grain size is essential for the existence of a superplastic behaviour. García-Bernal et al. [18] conducted a study to evaluate the high strain rate superplasticity behaviour during the high-temperature deformation of a continuous cast Al-Mg alloy, having reported that the generation of a fine grain structure and the breaking of cast structure led to a significant improvement in its ductility up to 800% at 530 °C and a strain rate of 3x10-2 s-1.

The fine-grained microstructure generated by FSP can also prevent fatigue crack initiation and propagation due to the barrier effect of grain boundaries. For example, Jana et al. [19] friction stir processed a cast Al-7Si-0.6Mg alloy, widely used for its good castability, mechanical properties and corrosion resistance, but characterized by poor fatigue properties. The authors succeeded to improve fatigue resistance by a factor of 15 at a stress ratio of $R=\sigma_{min}/\sigma_{max}=0$ due to a significant enhancement of ductility and a homogeneous redistribution of refined Si particles.

Intense plastic deformation and material mixing featured in the FSP of A356 aluminium casting also resulted in the significant breakage of primary aluminium dendrites and coarse Si particles, creating a homogenous distribution of Si particles in the aluminium matrix and eliminating casting porosity [7]. This led to a significant improvement of ductility and fatigue strength in 80%, proving that FSP can be used as a tool to locally modify the microstructures in regions experimenting high fatigue loading.

Friction surfacing of AA6082-T6 over AA2024-T3 evidenced a significant improvement of wear performance in about 25 %, compared to the consumable rod in as-received condition (Table 1). This enhancement in wear behaviour is also due to a finer equixial grain microstructure within the coating, compared to the rod anisotropic microstructure which is more prone to delamination under wear loads.

AA2024-T3 substrate plates exhibited the best tribological properties, presenting the lowest weight loss, frictional force and friction coefficient. This is most likely due to both its higher surface hardness and its lower ductility, which make this material less prone to suffer plastic deformation under abrasive wear, in comparison with the AA6082 coating and the rod in as-received condition. Due to the fine grain structure observed, the coatings present high frictional force and coefficient (10.9 N and 0.56, respectively).

Table 2. Weight loss due to wear (average values).

Material	Weight lost [mg]	Volume lost [m m³]	Volume rate [10^{-2} mm³/m]	Wear rate [mg/m]	First stage Run-in wear		Second stage Steady state wear	
					Frictional force [N]	Friction coefficient	Frictional force [N]	Friction coefficient
Substrate	12.6 ± 3	4.54	1.51	0.042	4.9 ± 0.97	0.25 ± 0.05	7.5 ± 0.33	0.38 ± 0.017
ARCR	30.2 ± 5	11.19	3.73	0.101	-	-	7.1 ± 1.14	0.36 ± 0.059
Coatings	23.2 ± 3	8.59	2.86	0.077	-	-	10.9 ± 0.58	0.56 ± 0.029

Surface Composites

FSP has been also investigated to produce layers of hard materials on soft substrates, as aluminium based alloys. Most of the published work is focused on the effect of processing parameters on surface characteristics and techniques to evaluate the performance of modified surfaces. Nevertheless, the reinforcing particles deposition method is relevant in terms of structural and chemical homogeneity and depth of the modified layer which influence the final surface performance. Different methods for depositing reinforced particles have been reported. A main reinforcing method consists of mixing reinforcing particles or powders with a volatile solvent such as methanol or a lacquer, in order to form a thin reinforcement layer, preventing reinforcing powders to escape. Another method consists of machining grooves in the substrate, pack these with reinforcing particles and process the zone with a non consumable FSP tool in a single pass or in multiple passes.

An enormous diversity of materials is used for surface reinforcements, the majority being hard ceramic particles as SiC, Al_2O_3 and AlN to improve surface properties as hardness, superplasticity, formability, corrosion and wear resistances.

The paricle size is relevant since small particles lead to higher concentration along bead surface and to smooth fraction gradients both in depth and along the direction parallel to the surface, while the thickness of the reinforced layer decreases with increasing particle size and is, typically, below 100 micron [20].

More recently, nanostructured layers have been produced and less common reinforcements were studied successfully. Two examples are: the incorporation of multi-walled carbon nanotubes (MWCNT) into a number of metallic materials as reinforcing fibres is a topic of recent interest due to the unique mechanical and physical properties of this material, namely very high tensile strengths [21]. FSP was tested to produce a composite of an aluminium alloy with MWCNT. Nanotubes were embedded in the stirred zone and the multi walled was retained. With tool rotational speeds of 1500 and 2500 rpm the distribution of nanotubes increased. Aiming at weight reduction of vehicles, FSP MWCNT/AZ31 surface composite were produced by Morisada et al. [22] and succeeded to disperse MWCNT into a AZ31 matrix. The microhardness increased to values of about 74 HV and the addition of MWCNT was seen to further promote grain refinement by FSP.

Another example is the incorporation of Nitinol (NiTi) that is a shape memory alloy with superelastic behavior and good biocompatibility. These alloys are widely used in orthodontics, but also in sensors and actuators. The possibility of incorporating wires, ribbons or powders into metallic matrixes opens up new applications for shape memory alloys. Studies report on the use of NiTi wires, but few have been made in the dispersion of NiTi powders in a metal matrix. Dixit et al. [23] produced a NiTi reinforced AA1100 composite using FSP and the particles were uniformly distributed. Good bonding with the matrix was achieved and no interfacial products were formed. The authors suggest that under adequate processing, the shape memory effect of NiTi particles can be used to induce residual stress in the parent matrix, of either compressive or tensile type. This study showed that samples had enhanced mechanical properties such as: Young modulus and micro hardness. A more recent work showed the possibility to introduce 1x2 mm ribbons of NiTi in AA1050 alloy by

FSP showing a good vibration and damping capacity of the composite [24].

Shafei-Zarghani et al. [25] used multiple-pass FSP to produce a superficial layer of uniformly distributed nano-sized Al2O3 particles into an AA6082 substrate. Hardness was increased three times over that of the base material. Wear testing revealed a significant resistance improvement. Researchers also found that the increase of the number of passes leads to more uniform alumina particle distributions with a significant increase of surface hardness. The nano-size Al2O3 powder was inserted inside a groove with 4 mm depth and 1 mm width, which was closed by a tool with a shoulder and no pin.

FRICTION SURFACING

Principles and Process Parameters

Friction surfacing (FS) was first patented in the 40's and is now well established as a solid state technology to produce metallic coatings. While FSP modifies the microstructure of a surface by simply deforming, recrystallize and homogenise the grain structure, FS modifies its chemistry. In friction surfacing a consumable rod under rotation is pressed under an axial load against the surface as depicted in Fig. 9. Heat generated in the initial friction contact promotes viscoplastic deformation at the tip of the rod. As the consumable travels along the substrate, the viscoplastic material at the vicinity of the rubbing interface flows into flash or is transferred over onto the substrate surface, while pressure and heat conditions triggers an inter diffusion process that soundly bonds the deposit. As the material undergoes a thermo-mechanical process, a fine grain microstructure is also produced by dynamic recrystallization.

Figure 9. Metallic coating of steel substrate by FS

Gandra et al [20] proposed a model for the global thermal and mechanical processes involved during friction surfacing based on the metallurgical transformations observed when depositing mild steel over mild steel and is shown in Fig.10.

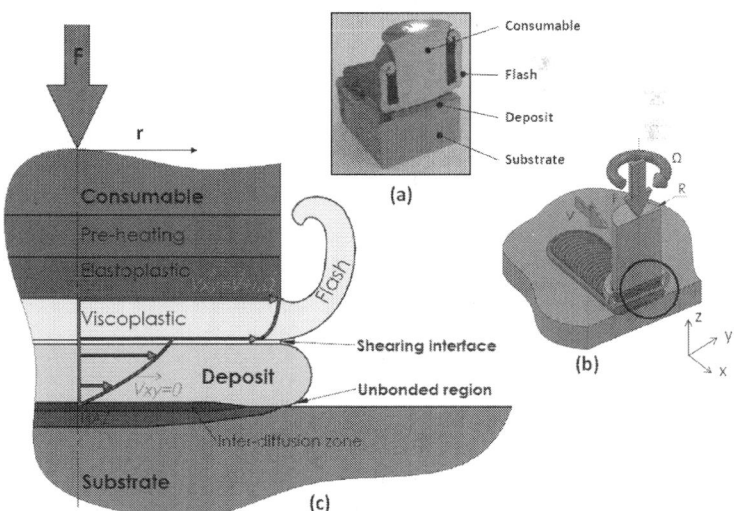

Figure 10. Thermo-mechanics of friction surfacing. (a) Sectioned consumable, (b) Process parameters and (c) Thermo-mechanical transformations and speed profile. Nomenclature: F – Forging force; Ω – rotation speed; v –travel speed; V_{xy} – rod tangential speed in-plan xy given by composition of rotation and travel movements [20]

The speed difference between the viscoplastic material, which is rotating along with the rod at v_{xy}, and the material effectively joined to the substrate ($v_{xy} = 0$), causes the deposit to detach from the consumable. This viscous shearing friction between the deposit and the consumable is the most significant heat source in the process.

Since the deposited material at the lower end is pressed without lateral confinement, it flows outside the consumable diameter, resulting into a revolving flash attached to the tip of the consumable rod and side unbounded regions adjacent to the deposit. Flash and unbonded regions play an important role as boundary conditions of temperature and pressure for the joining process.

Fig 11 shows typical material combinations tested using FS with successful results.

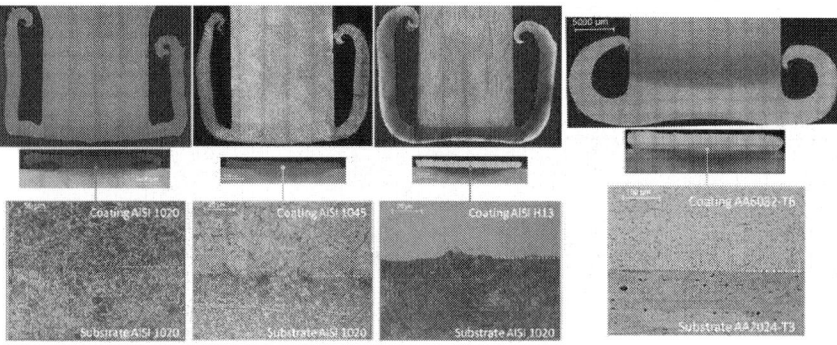

Figure 11. Different coatings/substrates combinations

The process allows the deposition of various dissimilar material combinations as the deposition of stainless steel, tool steel, copper or Inconel on mild steel substrates, as well as, stainless steel, mild steel and inconel consumables on aluminium substrates.

The influence of processing parameters on the deposit characteristics and bonding strength has been studied [26,27] aiming to correlate the resulting coating geometrical characteristics (thickness and bonded width) and mechanical performance with forging force, spindle and travel speeds. The increase of forging force improves the bond strength and reduces the coating thickness. The undercut region decreased when the forging force increased and the travel speed decreased. Higher ratios between the consumable rod feeding rate and the travel speed resulted in superior bonding quality. The applied load on the consumable rod was found to be essential to improve joining efficiency and to increase the deposition rate. Higher rotation or travel speeds were detrimental for the joining efficiency. Tilting the consumable rod along the travel direction proved to improve the joining efficiency up to 5 %. The material loss in flashes represented about 40 to 60 % of the total rod consumed, while unbonded regions were reduced to 8 % of the effective coating section in mild steel deposition. Friction surfacing was seen to require mechanical work between 2.5 and 5 kJ/g of deposited coating with deposition rates of 0.5 to 1.6 g/s, that is, deposition rates are higher than for laser cladding or plasma arc welding and the specific energy consumption lower than for other cladding processes.

In the friction surfacing of low carbon steel with tool steel H13 consumable rods, Rafi et al. [28] concluded that the coating width was strongly influenced by the rotation speed, while thickness was mostly determined by the travel speed.

This field of exploitation of producing aluminium coatings on aluminium based alloys is very promising. It was seen that friction surfacing enables intermediate mass deposition rates and higher energy efficiency in comparison with several mainstream laser and arc welding cladding processes. The required mechanical work varied between 2.5 and 5 kJ/g of deposited coating with deposition rates of 0.5 to 1.6 g/s. The forging force enhances joining quality while contributing to a higher overall coating efficiency. Faster travel and rotation speeds improved deposition rates and coating hardness, while decreasing energy consumption per unit of mass. Surface hardness increased up to 115 % compared to consumable rod. By adjusting a proper tilt angle, specific energy consumption drops, while slightly improving deposition rate and joining efficiency.

SUMMARY AND FUTURE TRENDS

Friction based processes comprise Friction Stir Processing (FSP) and Friction Surfacing (FS).

Friction stir processing is mostly used to locally eliminate casting defects and refine microstructures in selected locations, for property improvements and component performance enhancement. Aluminium and steel castings are amongst the most common components improved by this technology aiming at eliminating porosities, destroy solidification structures with inhomogeneous segregated phases, refine grain structures improving n-service performance.

The recent advances in adding reinforcing particles to manufacture surface alloys and metal matrix composites is a breakthrough in this technology opening new possibilities to manufacture composites nanostructured with tremendous properties.

Friction surfacing has been used in the production of long-life industrial blades, wear resistant components, anti-corrosion coatings and in the rehabilitation of worn or damaged parts such as, turbine blade tips and agricultural machinery. Other applications feature the hardfacing of valve seats with stellite and tools such as punches and drills.

Since the deposits result from severe viscoplastic deformation, friction surfacing presents some advantages over other coating technologies based on fusion welding or heat-spraying processes, that produce coarse microstructures and lead to intermetallics formation, thereby deteriorating the mechanical strength of the coatings. However, friction surfacing currently struggles with several technical and productivity issues which contribute to a limited range of engineering applications.

REFERENCES

1. Bunshah RF. Handbook of Deposition Technologies for Films and Coatings: Science, Technology and Applications. 2nd Edition. Noyes Publications; 1994.
2. Klopstock H, Neelands AR. Patent specification, An improved method of joining or welding metals; Ref. 572789; 1941.
3. Nicholas ED. Friction surfacing, ASM handbook: Vol 6. ASM International; 1993; p. 321–323.
4. Bedford GM, Vitanov VI, Voutchkov II. On the thermo-mechanical events during friction surfacing of high speed steels. Surface and Coatings Tecnhology 2001; 141 (1) 34-39.

REFERENCES

5. Thomas W. Friction Stir But Welding, International Patent Application N° PCT/GB92/02203 and GB Patent Application N° 9125978.8, US Patent N°5460317.
6. Mishra RS, Ma ZY. Friction stir welding and processing. Materials Science & Engineering R 2005; 50 (1-2) 1-78.
7. Nascimento F, Santos T, Vilaça P, Miranda RM, Quintino L. Microstructural modification and ductility enhancement of surfaces modified by FSP in aluminium alloys. Materials Science and Engineering A 2009; 506 (1-2) 16-22.
8. Chang ADu, XH, Huang JC, Achieving ultrafine grain size in Mg-Al-Zn alloy by friction stir processing. Scripta Materiallia 2007; 57 (3) 209-212.
9. Gandra J, Miranda R, Vilaca P, Velhinho A, Pamies-Teixeira J. Functionally graded materials produced by friction stir processing. Journal of Materials Processing Technology 2011; 211 (11) 1659-1668.
10. Porter DA, Easterling KE, Phase Transformations in Metals and Alloys 1992, CRC Press.
11. Gandra J, Miranda RM, Vilaça P. Effect of overlapping direction in multipass friction stir processing. Materials Science and Engineering A 2011; 528 (16-17) 5592–5599.
12. Nakata K, Kim YG, Fuji H, Tusumura T, Komazaki T. Improvement of mechanical properties of aluminium die casting alloy by multi-pass friction strir processing. Materials Science and Engineering A 2006; 437 (2) 274-280.
13. Santella ML, Engstrom T, Storjohann D, Pan TY. Effects of friction stir processing on mechanical properties of the cast aluminum alloys A319 and A356. Scripta Materialia 2005; 53 (2) 201-206;
14. Feng AH, Ma ZY. Enhanced mechanical properties of Mg-Al-Zn cast alloy via friction stir processing. Scripta Materiallia 2007; 56 (5) 397-400.
15. Charit I, Mishra RS, Low temperature superplasticity in a friction-stir-processed ultrafine grained Al-Zn-Mg-Sc alloy. Acta Materialia 2005; 53 (15) 4211-4223.
16. Liu FC, Ma ZY. Achieving exceptionally high superplasticity at high strain rates in a micrograined Al-Mg-Sc alloy produced by friction stir processing. Scripta Materialia 2008; 59 (15) 882-885.
17. Liu FC, Ma ZY, Chen LQ. Low-temperature superplasticity of Al-Mg-Sc alloy produced by friction stir processing. Scripta Materialia 2009; 60 (5) 968-971.
18. García-Bernal MA, Mishra RS, Verma R, Hernández-Silva D, High strain rate superplasticity in continuous cast Al-Mg alloys prepared via friction stir processing. Acta Materialia 2009; 60 (10) 850-853.

19. Jana S, Mishra RS, Baumann JB, Grant G. Effect of stress ratio on the fatigue behavior of a friction stir processed cast Al-Si-Mg alloy. Scripta Materialia 2009; 61 (10) 992-995.
20. Gandra J, Miranda RM, Vilaça P. Performance Analysis of Friction Surfacing. Journal of Materials Processing Technology 2012; 212 (8) 1676-1686.
21. Arora HS, Singh H, Dhindaw BK. Composite fabrication using friction stir processing - a review. International Journal of Advanced Manufacturing Technologies 2012; 61 (9-12) 1043–1055.
22. Morisada Y, Fujii H, Nagaoka T, Fukusumim M, MWCNTs/AZ31 surface composites fabricated by friction stir processing. Materials Science and Engineering A 2006; 419 (1-2) 344–348.
23. Dixit M, Newkirk JW, Mishra RS. Properties of friction stir-processed Al1100-NiTi composite. Scripta Materialia 2007; 56 (6) 541-544.
24. Mendes L. Production of aluminium based metal matrix composites reinforced with embedded NiTi by friction stir welding. MSc thesis, Universidade Nova de Lisboa, 2012
25. Shafei-Zarghani A, Kashani-Bozorg SF, Zarei-Hanzaki. Microstructures and mechanical properties of Al/Al2O3 surface nano-composite layer by friction stir processing. Materials Science and Engineering A 2009; 500 (1-2) 84-91.
26. Vitanov VI, Voutchkov II, Bedford GM. Decision support system to optimize the Frictec (friction surfacing) process. Journal of Materials Processing Technologies 2000; 107 (1-3) 236-242.
27. Vitanov VI, Javaid N, Stephenson DJ. Application of response surface methodology for the optimisation of micro friction surfacing process. Surfaces and Coatings Technology 2010; 204 (21-22) 3501-3508.
28. Khalid Rafi H, Janaki Ram GD, Phanikumar G, Prasad Rao K. Microstructural evolution during friction surfacing of tool steel H13. Materials and Design 2011; 32 (1) 82–87.

CITATION

R. M. Miranda, J. Gandra and P. Vilaça (2013). Surface Modification by Friction Based Processes, Modern Surface Engineering Treatments, Dr. M. Aliofkhazraei (Ed.), ISBN: 978-953-51-1149-8, InTech, DOI: 10.5772/55986.

CHAPTER 8

Micro/Nano-Mechanical Sensors and Actuators Based on Soi-Mems Technology

Dzung Viet Dao, Koichi Nakamura, Tung Thanh Bui and Susumu Sugiyama

Research Institute for Nanomachine System Technology, Ritsumeikan University, 1-1-1 NojiHigashi

ABSTRACT

MEMS (micro-electro-mechanical systems) technology has undergone almost 40 years of development, with significant technology advancement and successful commercialization of single-functional MEMS devices, such as pressure sensors, accelerometers, gyroscopes, microphones, micro-mirrors, etc. In this context of MEMS technology, this paper introduces our studies and developments of novel micro/nano-mechanical sensors and actuators based on silicon- on-insulator (SOI)-MEMS technology, as well as fundamental research on piezoresistive effects in single-crystal silicon nanowires (SiNWs). In the first area, novel mechanical sensors, such as 6-DOF micro-force moment sensors, multi-axis inertial sensors and micro-electrostatic actuators developed with SOI-MEMS technology will be presented. In the second area, we have combined atomic-level simulation and experimental evaluation methods to explain the giant piezoresistive effect in single crystalline SiNWs along different crystallographic orientations. This discovery is significant for developing more highly sensitive and miniaturized mechanical sensors in the near future.

INTRODUCTION

MEMS and NEMS stand for, respectively, micro-electromechanical systems and nano-electro-mechanical systems, which relate to micro/nano-electro-mechanical integrated devices fabricated by the extension of microelectronic fabrication technology, e.g. photolithography, thin film deposition and etching, with high accuracy and high throughput. While microelectronic devices are solid and mechanically immovable, MEMS devices have movable 3D microstructures, e.g. micro-cantilevers, micro-beams, membranes, etc. However, the significance of MEMS/NEMS is not only in mechanical motion, but also in miniaturization, multifunctional integration and mass production. Compared to LSI devices, which deal only with electrical signals, MEMS devices relate to conversion and integration of a wide variety of signal types, such as physical (electrical, mechanical, thermal, optical, etc), chemical and biological signals. Generally, silicon MEMS technology offers the possibility of low-cost, high- performance, and miniaturized multifunctional-integrated devices for use in a wide range of consumer and industrial applications, such as in the automotive, biomedical and telecommunication industries, defense, and so on.

The first ideas on MEMS/NEMS were proposed by the 1965 Nobel laureate Richard Feynman in 1959 [1]. However, MEMS devices were not demonstrated until the late 1960 s and early 1970 s when some silicon micro-devices and structures, such as gas chromatography [2, 3], pressure sensors [4–6], force sensors [7], resonators [8], micromirrors [9], ink-jet nozzles [10], etc, were successfully developed. This landmark of MEMS device development was made possible because of the previous studies and success of so-called bulk silicon micromachining technology in the late 1960 s and, subsequently, surface micromachining technology in 1986–1988. The former mainly features in wet isotropic etching [11, 12], wet anisotropic etching [13, 14], dry anisotropic etching (e.g. deep reactive-ion etching (D-RIE) [15, 16]), as well as wafer bonding techniques [17, 18], while the latter relates to selectively etching thin sacrificial layers underlying etch-resistant thin films from a multilayer sandwich of patterned thin films [19].

After decades of extensive study in both microfabrication techniques and material properties [20], silicon was considered to be the most appropriate and useful material for MEMS technology. Besides the fact that silicon is an excellent electronic material used in most integrated circuits, it also has excellent mechanical properties: high elasticity, stiffness, fracture and fatigue strengths, thermal conductivity, thermal stability, etc. Furthermore,

INTRODUCTION

silicon has many properties extremely suitable for highly sensitive integrated sensor applications, such as piezoresistive effect [21], thermoresistive effect, photovoltaic effect, and so on.

Over the past 40 years of research and development of MEMS, there have been significant changes in the field: huge improvements in the technology, extensive infrastructure in place, significant designs for commercialization (such as pressure sensors, accelerometers, gyroscopes, flow sensors, ink jet print heads, microphones, resonators, digital mirror displays, optical switches, etc), The MEMS devices market had reached US$7.1 billion in 2007, US$6.8 billion in 2008, and was estimated to reach US$6.9 billion and US$49 billion in 2009 and 2020, respectively [22]. MEMS technology has attracted the attention of worldwide research institutions to develop new industry, new applications and new markets. While MEMS have been successfully developed and commercialized, NEMS technology is still being fundamentally studied to elucidate new properties and effects of materials at the nanoscale to further enhance the performance and widen the application of micro-integrated devices, while reducing the overall size. Mechanical, electrical, optical, chemical and biological properties, and various sensing effects of nanomaterials such as nanodots, nanowires and nanosheets, are being investigated.

In this paper, we present our recent study on micro-devices based on SOI-MEMS, as well as using MEMS as a platform for studying the piezoresistive effect in single crystalline silicon nanowires. All devices have been fabricated using SOI (silicon-on-insulator) wafers. The term SOI refers to a silicon substrate with a single crystalline silicon device layer lying on a buried oxide layer which is set on top of a silicon substrate. This structure is widely used in MEMS because it provides reliable electrical insulation, excellent etching stop and sacrificial layer functions; therefore, it increases the fabrication accuracy, process simplicity and device performance. MEMS industry is boosting SOI wafer demand thanks to the market dynamism of products such as accelerometers or gyroscopes, which are now widely used in numerous consumer products and industry.

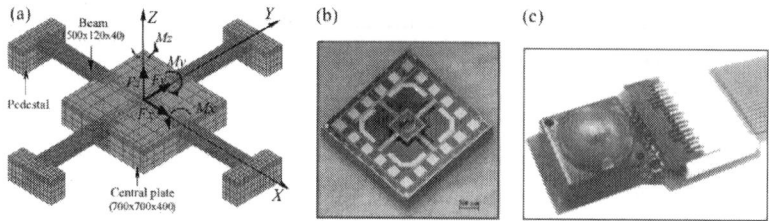

Figure 1. 6-DOF force moment sensor (a, b) and tactile sensor for robotic application (c). (a) Simulation model, (b) fabricated force sensor chip and (c) tactile sensor.

MICRO-MECHANICAL SENSORS AND ACTUATOR

Here we introduce our typical SOI-MEMS micro-mechanical sensors and actuators, including a six-degree-of-freedom (6-DOF) force moment sensor, a 3-DOF micro-accelerometer, a dual-axis gyroscope and an electrostatically driven micro-transportation system.

6-DOF force moment sensor

The sensing chip was designed to be able to simultaneously detect three components of force and three components of moment in three orthogonal directions (figure 1). Two-terminal and four-terminal p-type silicon piezoresistors were created by thermal diffusion of boron ions to suitable places on the surface of a crossbeam-shaped structure of (111) n-type silicon. The total number of piezoresistors is 18, much fewer than the previous piezoresistive-based 6-DOF force moment sensors known to the authors [23]. When a force or moment is applied to the sensor, the beams will be deformed; consequently, stresses will be induced at the piezoresistors, leading to a change of their resistance, thus changing the output of the corresponding bridge circuits. The sensor was fabricated based on IC compatible processes, e.g. photolithography, impurity diffusion, thin film deposition and etching, and silicon bulk micromachining technology, e.g. the deep-RIE etching process. The schematic of the fabrication process is shown in figure 2. The fabricated chip has dimensions of $3 \times 3 \times 0.5$ mm^3 as shown in figure 1(b). The sensor has been applied in robotics as fingertip tactile sensors (figure 1(c))

[24], and in hydraulics to measure the fluid force acting on small particles in water flow [25].

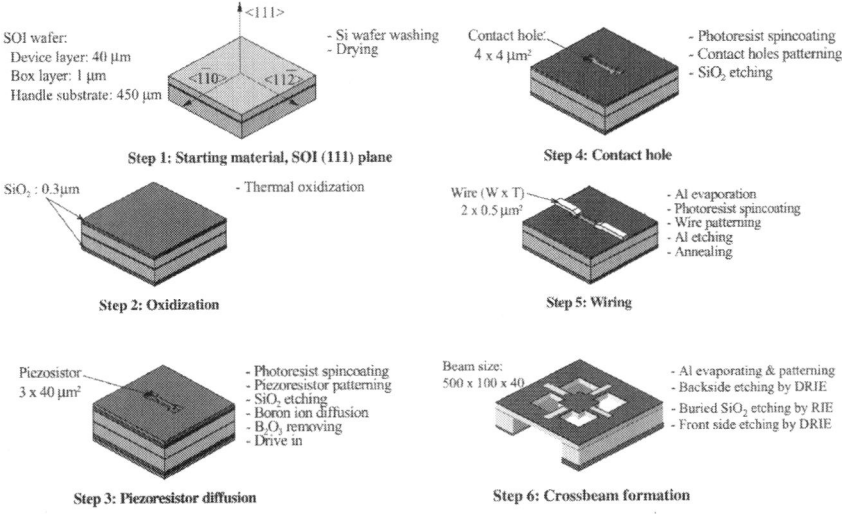

Figure 2. Fabrication process of the 6-DOF force moment sensor.

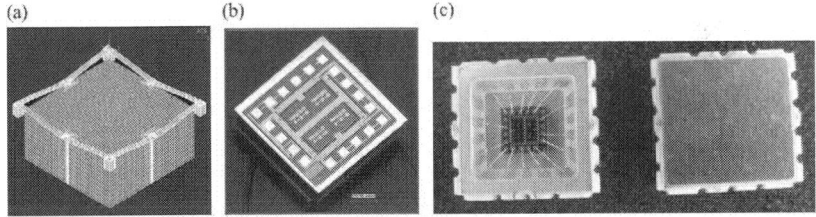

Figure 3. Simulation model (a) and photos of the fabricated 3-DOF micro-accelerometers (b, c).

Multi-axis Micro-accelerometer

Inertial sensors refer to accelerometers and angular rate sensors (or gyroscopes). Several kinds of micro-accelerometers, such as piezoresistive accelerometers [26, 28], thermo-resistive accelerometers [27], have been developed. A recently developed accelerometer is shown in figure 3 [28]. This accelerometer can independently detect three components of linear acceleration along three orthogonal directions. The detecting principle is based on piezoresistive effects in single-crystal silicon. The sensing chip is made of single-crystal Si, with a seismic mass being suspended at the middle of the four surrounding beams, which are themselves fixed to the 'rigid' frame at the ends. Si piezoresistors are formed by diffusing boron ions to suitable places on the surface of the n-type silicon sensing beams. When an external acceleration is applied to the sensor, the seismic mass is displaced due to the inertial force. This movement of the seismic mass deforms the beams. As a result, the resistance of the Si piezoresistors is changed due to the piezoresistive effect. The change of resistance is converted to an output voltage by Wheatstone bridges. The fabrication process is illustrated schematically in figure 4, and an SOI wafer was used. The dimensions of the accelerometer chip are 1 mm × 1 mm × 0.45 mm (L × W × T). The sensitivities to the X-, Y- and Z-axis are 30, 30 and 23µV g^{-1} , respectively, and cross-axis sensitivity is less than 5.5%. Thanks to the small size and high performance, the sensor can be applied to portable electronic devices, such as mobile phones or cameras to stabilize or automatically find the image orientation, or to remote game controllers, health monitoring sensors, etc

Dual-axis angular rate sensor (gyroscope)

Figure 5 shows our recently developed 2-DOF convective angular rate sensor (gyroscope) [29, 30]. The sensor can detect two components of angular velocity based on the thermo-resistive effect of the low-doped silicon. The working principle is illustrated in figure 5(a). The gas flow generated by the piezoelectric diaphragm pump flows to the nozzle orifice and creates a jet flow in the chamber. The gas flows toward the sensing element, and goes through the symmetric center of the four thermistors. As an angular rate is applied, the gas flow is deflected (figure 5(a)) due to the Coriolis effect. This deflection causes differential cooling between opposing thermistors and, as a result, the resistances of the two thermistors change in opposite directions. A Wheatstone bridge is used to convert these resistance changes to output voltage. Therefore, angular rates ω_x around the X-axis and ωy around the Y-axis can be measured independently. The sensing chip is fabricated from the SOI wafer by applying a fabrication process similar to that shown in figure 2. The gyroscope was packaged at Tamagawa Seiki Corp. (figure 5(b)). The sensitivities of the gyroscope for the X and Y-axis are 0.082 and 0.078 mV deg^{-1} sec^{-1} , respectively. The cross-sensitivities between the two input axes are less than 0.26%; the nonlinearity was smaller than 0.5% F.S in the range of ±200 deg sec^{-1} . Several applications of this high-performance gyroscope mentioned here include automotive, ship stabilization systems and aerospace.

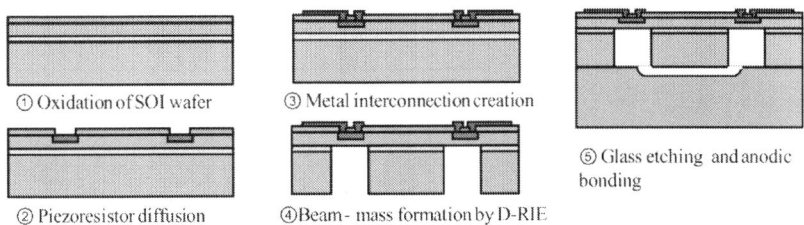

① Oxidation of SOI wafer
② Piezoresistor diffusion
③ Metal interconnection creation
④ Beam- mass formation by D-RIE
⑤ Glass etching and anodic bonding

Figure 4. Fabrication process of the 3-DOF Si micro-accelerometer.

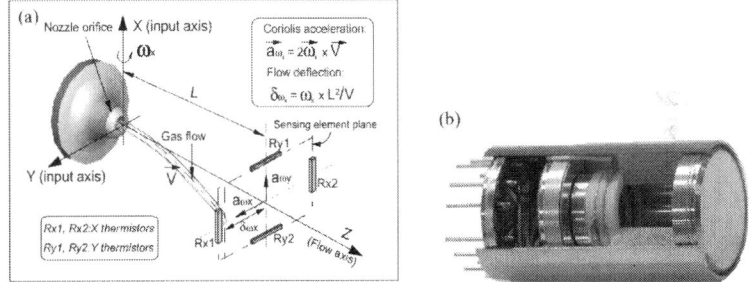

Figure 5. 2-DOF convective angular rate sensor [29, 30]. (a) Working principle and (b) cut-away view of the gyroscope.

Electrostatic Micro-actuator

Silicon comb-drive electrostatic actuators are among the most frequently utilized in MEMS since they were first reported about 20 years ago [31]. The advantages of these actuators include the simple and mass fabrication process, accuracy, easy control and large displacement. In this paper, we propose novel ratcheting structures which allow converting reciprocating displacement of electrostatic comb-drive actuators to continuous, unidirectional, straight and turning movement of micrometer-scale objects. The working principle of the micro-transportation system is shown schematically in figure 6. Force induced from the electrostatic comb-drive actuator pushes the ratchet racks inward, causing the driving wings to move inward. As a result, the container (moved object) is moved forward. When the force is removed, the container will not move backward due to the ratchet mechanism of the anti-reverse wings. Figure 7 shows the silicon micro-transportation system [32, 33] made from SOI wafer by using D-RIE silicon etching and SiO_2 etching processes. The velocity of the object can be controlled by driving voltage and frequency. The maximum velocity was about 200μm s^{-1} at 100 V and 20 Hz driving voltage. This miniaturized transmission system can be used to transport small objects, such as biomedical samples, micro/nano-particles and carbon nanotubes (CNTs) from one place to another.

Another transmission system based on an electrostatic actuator and ratchet mechanism, called a gearing transmission system, was developed recently as shown in figure 8 [34]. The reciprocal movement of rotational comb actuators is converted to the unidirectional rotation of the gears through a driving mechanism as shown in the right SEM image of figure 8. The system was fabricated from SOI wafer using D-RIE and HF vapor etching processes. The angular velocity was changeable depending on the frequency of the driving voltage. The maximum angular velocity was about 40 deg s^{-1} at 80 V and 50 Hz driving voltage. The gearing system can be used in micro-motors, micro-watches or in other high- precision instruments.

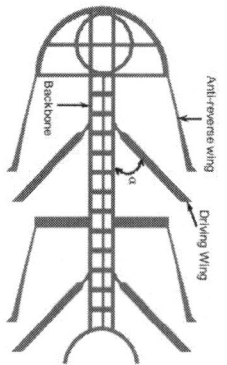 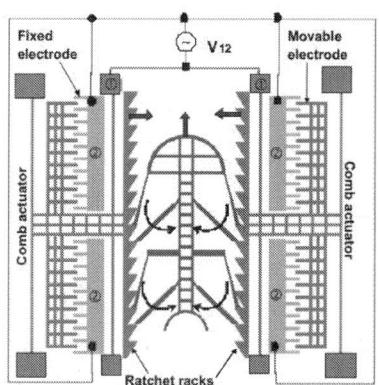

Figure 6. Structure of a micro-container (left), and a schematic view showing the working principle of the system (right). Pitch and height of ratchet teeth are 10μm and 6μm, respectively.

 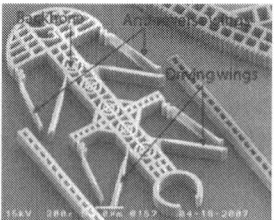

Figure 7. SEM images of a micro-transportation system (left) and micro-containers (right). Length, width and thickness of the container are 450μm, 250μm and 30μm, respectively.

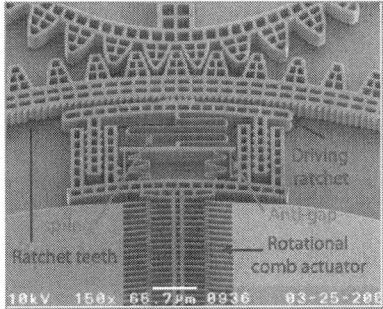

Figure 8. SEM images of a micro-gearing transmission system (left) and close-up image of the driving mechanism of the gearing transmission system (right). Diameters of the driving and driven gears are 2.4 mm and 0.8μm, respectively.

FUNDAMENTAL STUDY OF THE PIEZORESISTIVE EFFECT IN SILICON NANOWIRE FOR MECHANICAL SENSING

In this section, we present our theoretical and experimental investigations of the piezoresistive effect of single crystalline silicon nanowires (SiNW), as well as a promising application of SiNW in the development of higher sensitive and miniaturized mechanical sensors in the near future. Firstly, the piezoresistive effect in SiNW is theoretically studied using atomic-level simulation, that is the first-principles calculation method, and then it is experimentally measured on the SiNW sample fabricated using electron-beam direct writing and RIE. Finally, the fabrication process of an ultra-small accelerometer having SiNW as sensing element is presented.

First-principles Simulation

We have performed first-principles calculations based on density functional theory [35] with the generalized-gradient approximation (GGA) method [36]. We adopted the three-dimensional supercell approximation technique with norm-conserving pseudopotentials prepared according to the Hamann method [37]. The cutoff energy for wave functions of electrons with plane-wave expansion was set at 30 Rydberg (408 eV).

Si nanowire models were composed of a fragment of optimized bulk Si with a one-dimensional periodic boundary, and all dangling bonds at the wire wall were terminated with H atoms. Figure 9 shows $Si_{89}H_{44}\langle 001\rangle$ (wire diameter 2R = 2.21 nm) and $Si_{80}H_{28}\langle 110\rangle$ (2R = 2.58 nm) Si nanowire models, where the longitudinal directions with one-dimensional periodic boundary are respectively set to [001] and [110].

A periodic boundary condition along the transverse directions, or perpendicular directions to the wire axis in the three-dimensional supercell, was given by inserting sufficient space between H-terminated Si nanowires with two large cell parameters perpendicular to the wire axis. On the condition that sufficient space in the Si nanowire model disturbs the interaction between the Si nanowires, the electronic band structure of the Si nanowire model can be reduced to one dimension, which is dependent on only one reciprocal coordinate, k_z. The effect of uniaxial tensile strain on structure was represented by partial optimization of bulk Si with a fixed lattice constant along the tensile direction.

The conductive properties of Si nanowire models have been discussed in terms of the band structures. The electrical conductivity G or the electrical resistivity ρ can be represented in terms of carrier density and effective mass. We have introduced the band carrier densities and their corresponding effective masses for each energy subband, and the conductivity has been added up over all valence-band (VB) subbands and conduction-band (CB) ones as follows:

$$G = \frac{1}{\rho} = e^2 \left(\sum_{j \in CB} \frac{n_j \tau_{e,j}}{m^*_{e,j}} + \sum_{j \in VB} \frac{p_j \tau_{h,j}}{m^*_{h,j}} \right), \tag{1}$$

where n_j is the j_{th} CB carrier-electron density, p_j is the j_{th} VB hole density, $m^*_{e,j}$ and $m^*_{h,j}$ are the band effective masses, $\tau_{e,j}$ and $\tau_{h,j}$ are the relaxation times and e_2 is the square of the absolute value of the elementary electric charge. Subscripts e and h, respectively, denote electron and hole carriers. In actual n- or p-doped Si nanowires, the total number of carriers per unit cell, δ, must be less than 1 by a few orders. Under the condition that a small amount of the carrier occupation does not cause significant changes in the band structure, δ can be given by an appropriate shift of the Fermi energy E_F as follows,

$$\delta = \sum_{j \in CB} n_j V = 2 \sum_{j \in CB} \sum_{k_z} w_{k_z} \left\{ \exp\left(\frac{E_{j,k_z} - E_F}{k_B T} \right) + 1 \right\}^{-1} \tag{2}$$

in the n-type carrier occupation, or

$$\delta = \sum_{j \in VB} p_j V = 2 \sum_{j \in VB} \sum_{k_z} w_{k_z} \left\{ \exp\left(-\frac{E_{j,k_z} - E_F}{k_B T}\right) + 1 \right\}^{-1}$$

(3)

in the p-type carrier occupation with temperature T, where E_{j,k_z} is the intrinsic-semiconductor-state band energy of the j_{th} band at the k_z point, w_{kz} is the k-point weight for k_z, V is the volume of Si nanowire in the unit cell and k_B is the Boltzmann constant. In practice, we first have to set δ to an appropriate constant such as 10^{-n} (n = 2, 3 and 4), and then E_F in n-type and p-type carrier occupations can be solved according to equations (2) and (3), respectively [38–41].

The effective mass is generally given by a 3 × 3 tensor, but it can be defined simply as a scalar for one-dimensional SiNW models [38, 39]:

$$m_j^* = \pm \hbar^2 \left(\partial^2 E_j / \partial k_z^2\right)^{-1}$$

(4)

at the maximum or minimum of band energy E_j $\hbar$ is Planck's constant divided by 2π. On the right-hand side, a positive sign is adopted for carrier electrons and a negative sign for holes. For the relaxation time, we have adopted the approximation that all of the band relaxation times are equal and constant regardless of stress [37–40].

The $Si_{89}H_{44}h001i$ model has four VB subbands in the vicinity of the VB top as shown in figure 10. The highest VB subband of those without stress has a small hole effective mass, called the light-hole band, and by contrast, two of the second highest VB subbands in degeneracy have a larger hole effective mass, called the heavy-hole bands. The uniaxial tensile stress in the [001] longitudinal direction causes a sharp drop in the band energy of the light-hole band, leading to the alternation of the order of band energy levels between the light-hole band and the heavy-hole bands, and then most of the holes will be redistributed to the heavy-hole bands where hole effective mass is markedly raised due to the longitudinal tensile stress. This sudden change in the hole occupation with the increase in effective mass will bring a significant decrease in the hole conductivity. As a result, 1% longitudinal strain for the p-doped $Si_{89}H_{44}h001i$ model, i.e., the H-terminated h001i SiNW model reduced the conductivity by as much as 66% in our calculation.

The longitudinal piezoresistance coefficient π_l and transverse one π_t are given by

$$\pi_l = \Delta \rho_l / \rho_0 \sigma_l, \quad \pi_t = \Delta \rho_t / \rho_0 \sigma_t, \tag{5}$$

where $\sigma_{l,t}$ are the longitudinal and transverse tensile stresses, which have been represented by a linear-response approximation according to the classical Hooke's law,

$$\sigma_l = Y_l \varepsilon_l, \quad \sigma_t = Y_t \varepsilon_t \tag{6}$$

with Young's modulus $Y_{l,t}$ and tensile strain $\varepsilon_{l,t}$, $\rho 0$ is the resistivity along the wire axis without stress, and $1\rho l,t$ are variations in $\rho 0$ due to $\sigma_{l,t}$. Young's modulus of nanoscale low-dimensional Si materials [42] is somewhat smaller than that of bulk Si. According to the first-principles total energy calculation of tensile-strained models, we assumed $Y_{l,t}$ to be 25% of the experimental values of Young's modulus of bulk Si.

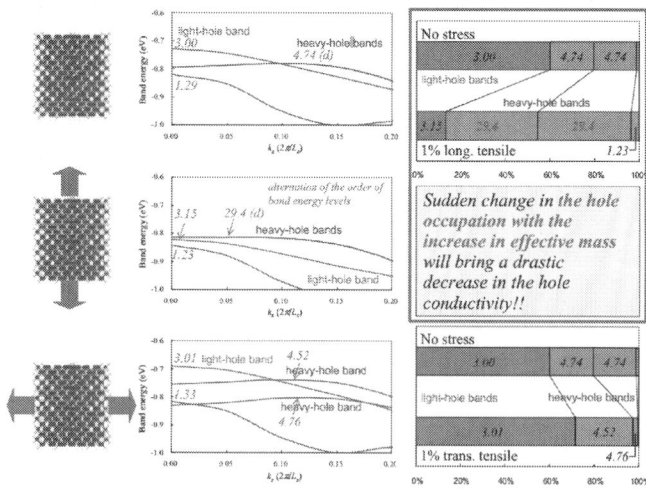

Figure 10. Variations of valence-band diagrams, occupation ratios and effective masses due to 1% longitudinal and transverse tensile strains for the Si89H44h001i SiNW model.

FUNDAMENTAL STUDY OF THE PIEZORESISTIVE EFFECT IN SILICON NANOWIRE FOR MECHANICAL SENSING

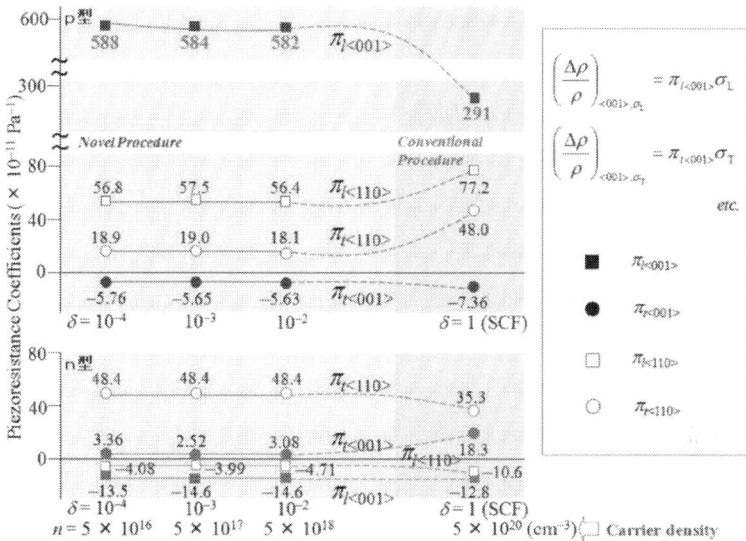

Figure 11. Piezoresistance coefficients of SiNW models with respect to carrier concentration N.

Figure 11 summarizes calculation results of the piezoresistance coefficients with respect to δ. Obviously, the values of $(\pi_{l\langle 001\rangle})_p$, longitudinal piezoresistance coefficients for the p-doped Si89H44h001i model, are by far the largest in figure 11. We have obtained 588 × 10^{-11} Pa^{-1} for $(\pi_{l\langle 001\rangle})_p$ with δ = 10^{-4}, i.e. two orders larger than that of the bulk silicon, and it is expected that p-doped h001i SiNW will have giant longitudinal piezoresistivity.

Measurement of the Piezoresistive Effect in SiNW

A giant piezoresistive effect has been theoretically found in single crystalline SiNWs using the first-principles calculation method. It is necessary to verify these results experimentally. Accordingly, SiNWs with a length of 2μm, thickness of 35 nm and width ranges from 35 nm to 480 nm have been fabricated from SIMOX (separation by implanted oxygen) SOI wafer using electron beam (EB) direct writing and RIE etching [43, 44]. Figures 12 (a) and (b) shows SEM images of the 35 nm width SiNWs array.

The width dependence of the longitudinal piezoresistive coefficient $\pi_{l\langle 110\rangle}$ along the h110i crystallographic orientation at doping concentration of 1.2 × 10^{18} atoms cm^{-3} of SiNWs has been measured (figure 12(c)). The results show a strong increase of the longitudinal piezoresistive coefficient when the width of the SiNW becomes smaller than 150 nm. The increase of $\pi_{l\langle 110\rangle}$ is up to 62.5% when the SiNW's width decreases to 35 nm. This enhancement of $\pi_{l\langle 110\rangle}$ can be qualitatively interpreted by introducing the one-dimensional (1D) hole transfer and the hole conduction mass shift mechanisms based on the 1D hole transport system, which might be expected to be induced in a nanowire p-type piezoresistor. This enhancement of the longitudinal piezoresistive coefficient is very significant for sensitivity improvement of piezoresistive-based ultra-small mechanical sensors, one of which will be presented in section 3.3.

Ultra-small Accelerometer Based on SiNW

The first demonstration of applying SiNWs as nanoscale piezoresistors to mechanical sensing was carried out. An ultra-small 3-DOF accelerometer utilizing Si nanowires as nanoscale piezoresistors was designed and fabricated. The model of the sensing chip is shown in figures 13(a) and (b). The seismic mass is suspended on four surrounding sensing beams, which are connected to the frame at the ends. Si nanowires are placed near the fixed ends on the surface of the sensing beams. When acceleration is applied to the sensor, the seismic mass will be moved due to the inertial force. This movement deforms the beams; as a result, the resistance of the nanowires changes. The change of resistance is converted to a voltage change by a Wheatstone bridge. The overall size of the accelerometer chip is $500 \times 500 \times 400 \mu m^3$, ($L \times W \times T$). This accelerometer has been made from multi-layer SIMOX SOI wafer and fabricated by nano/ micromachining technology, including EB lithography and RIE silicon etching [45]. The SiNW piezoresistor has dimensions of 128 nm × 50 nm (W × T). The resistance of these piezoresistors was measured to be 20 kΩ, and the calculated sensitivity for each axis is about 50μV G^{-1} , and the resolution is 30 mG. (G is the gravitational acceleration, i.e. = 9.8 m s^{-2}). Accordingly, by using SiNWs as piezoresistors, we can reduce the sensor's size while increasing the sensitivity. Smaller chip size means more chips per wafer, higher productivity, and therefore lower cost per chip. Ultra-small accelerometers are important for portable devices, such as camcorders, mobile phones, navigation systems, entertainment devices, and so on.

CONCLUSIONS

This paper has briefly described progress in micro/nanoelectromechanical systems (MNEMS) technology, as well as several research achievements of our group on SOI-MEMS mechanical sensors and electrostatic actuators. SOI wafers have been used because the buried oxide layer plays very important roles; for example, it is a reliable electrical insulation layer, an excellent etching stop layer and an ideal sacrificial etching layer. Therefore, it has become the principal substrate for MEMS/NEMS device fabrication. Furthermore, the cost per unit of SOI wafer is decreasing due to the rapid development of silicon wafer fabrication technology.

Over the past 40 years, single functional MEMS devices have developed rapidly and their consumer applications have spread widely in industries such as automotive, telecommunication, biomedical, security and defense, etc. Recently, MEMS have also become good platforms for studying nanotechnology. MEMS technology has lagged behind the IC industry, which came into production in the mid-1960s, by about 15 to 20 years. However, with significant technology improvement, strong development of MEMS infrastructure and strong demand for higher performance, multifunction integrated, low-cost and miniaturized devices, MEMS technology will continue to play an important role in industry and consumer applications. Future trends of MEMS devices will be toward multifunctional integration, nanoscale integration, integration with electronics, and new MEMS materials to create new and smart devices for not only traditional applications but also new purposes such as Environmental-MEMS, Energy-MEMS, Safety-MEMS, Health-MEMS, and so on.

ACKNOWLEDGMENTS

This study was partially supported by the Ministry of Education, Culture, Sports, Science and Technology of Japan under the Grant-in-Aid for Young Scientists (No. 21710141-0001, 2009–2010).

REFERENCES

1. Feynman R P 1992 J. Microelectromech. Syst. 1 60.
2. Terry S C 1975 A gas chromatography system fabricated on a silicon wafer using integrated circuit technology PhD Dissertation Department of Electrical Engineering, Stanford University, CA, USA
3. Terry S C et al 1979 IEEE Trans. Electron Devices 26 1880
4. Samaun, Wise K D and Angell J B 1973 IEEE Trans. Biomed. Eng. 20 101
5. Frobenius W D, Sanderson A C and Nathanson H C 1973 IEEE Trans. Biomed. Eng. 20 312
6. Ko W H, Hynecek J and Boettcher S F 1979 IEEE Trans. Electron Devices 26 1896
7. Tuan H C, Yanacopoulos J S and Nunn T A 1975 Stanford Univ. Electron. Res. Rev. p 102
8. Wilfinger R J, Bardell P H and Chhabra D S 1968 IBM J. Res. Dev. 12 113
9. Petersen K E 1980 IBM J. Res. Dev. 24 631
10. Bassous E, Taub H H and Kuhn L 1977 Appl. Phys. Lett. 31 135
11. Robbins H and Schwartz B 1959 J. Electrochem. Soc. 106 505
12. Bogenschlltz A F, Krusemark W, Cherer K-H L and Mussinger W 1967 J. Electrochem. Soc. 114 970
13. Finne R M and Klein D L 1967 J. Electrochem. Soc. 114 965
14. Lee D B 1969 J. Appl. Phys. 40 4569
15. Laermer F and Schilp A (Robert Bosch GmbH) 1996 US Patent No. 5501893
16. Pandhumsoporn T, Feldbaum M, Gadgil P, Puech M and Maquin P 1996 Proc. SPIE 2879 94–102
17. Wallis G and Pomerantz D I 1969 J. Appl. Phys. 40 3946
18. Denee P B 1969 J. Appl. Phys. 40 5396
19. Roger T H 1988 J. Vac. Sci. Technol. B 6 1809
20. Petersen K E 1982 Proc. IEEE 70 420
21. Smith C S 1954 Phys. Rev. 94 42
22. Eloy J C 2009 Status of the MEMS Industry—2009 Report Yole Développement Corp http://www.yole.fr/pagesAn/ products/mis.asp
23. Dao D V, Toriyama T, Wells J and Sugiyama S 2002 IEEJ Trans. Sens. Micromach. 122 35
24. Dao D V, Toriyama T, Wells J and Sugiyama S 2007 IEEJ Trans. Sens. Micromach. 127 177
25. Dao D V, Toriyama T, Wells J and Sugiyama S 2003 Sensors. Mater. 15 113
26. Amarasinghe R, Dao D V, Toriyama T and Sugiyama S 2005 J. Micromech. Microeng. 15 1745
27. Dau V T, Dao D V, Hayashida M, Dinh T X and Sugiyama S 2006 IEEJ Trans. Sens. Micromach. 126 190

REFERENCES

28. Bui T T et al 2008 IEEJ Trans. Sensor. Micromach. 128 235
29. Dao D V et al 2007 J. Micro Electromech. Sys. 16 950
30. Dau V T, Dao D V, Shiozawa T and Sugiyama S 2008 J. IEEE Sensor 8 1530
31. Tang W C, Nguyen T H and Howe R T 1989 Laterally driven polysilicon resonant microstructures Tech. Dig. IEEE Micro Electro Mechanical Systems Workshop pp 53–61
32. Pham P H, Dao D V and Sugiyama S 2007 J. Micromech. Microeng. 17 2125
33. Dao D V, Pham P H and Sugiyama S 2008 A fully functional electrostatic micro transportation system with strider-lick movement of micro containers Tech. Dig. IEEE MEMS08 (Tucson, AZ, USA) pp 50–3
34. Dao D V, Pham P H, Amaya S and Sugiyama S 2008 Micro ratcheting transmission systems based on electrostatic actuator IEEE Int. Symp. on Micro-Nano Mechatronics and Human Science (Nagoya, Japan) pp 17–20
35. Hohenberg P and Kohn W 1964 Phys. Rev. B 136 864
36. Perdew J P, Burke K and Ernzerhof M 1996 Phys. Rev. Lett. 77 3865
37. Hamann D R 1989 Phys. Rev. B 40 2980
38. Nakamura K, Isono Y and Toriyama T 2008 Japan. J. Appl. Phys. 47 5132
39. Nakamura K, Isono Y, Toriyama T and Sugiyama S 2009 Japan. J. Appl. Phys. 48 06FG09
40. Nakamura K, Isono Y, Toriyama T and Sugiyama S 2009 Phys. Rev. B 80 045205
41. Nakamura K, Isono Y, Toriyama T and Sugiyama S 2010 IEE J. Trans. Electr. Electron. Eng. 5 at press
42. Fedorchenko A I, Wang A B and Cheng H H 2009 Appl. Phys. Lett. 94 152111
43. Bui T T, Dao D V, Toriyama T and Sugiyama S 2009 Evaluation of the piezoresistive effect in single crystalline silicon nanowires Proc. IEEE Sensors 2009 (Christchurch, New Zealand) pp 41–4
44. Nakamura K and Dao D V et al 2009 Piezoresistive effect in silicon nanowires—a comprehensive analysis based on first-principles calculations Int. Symp. on Micro-Nano Mechatronics and Human Science (Nagoya, Japan) pp 462–6
45. Dao D V, Toriyama T and Sugiyama S 2004 Noise and frequency analyses of a miniaturized 3-DOF accelerometer utilizing silicon nanowire piezoresistors Proc. IEEE Sensors 2004 (Vienna, Austria) pp 1464–7

CITATION

Dzung Viet Dao, Koichi Nakamura, Tung Thanh Bui and Susumu Sugiyama, Micro/nano-mechanical sensors and actuators based on SOI-MEMS technology, doi:10.1088/2043-6254/1/1/013001.

Index

A
Architectural design research, 55, 56

B
Bearing arrangement, 121, 122, 124, 127, 132, 140, 141, 142, 145, 149, 150, 153, 154, 155, 167
Bevel gear, 22

C
Chemical vapour deposition (CVD), 223
Complex analysis, 119
Computer-aided architectural, 56
Contextual variation, 63
Cryogenic pumps, 169, 179
Cryptography, 55, 69
Cutting process, 119, 121, 126
Cyclical relationships, 58

D
Data acquisition and control (DAC), 7
Densification, 224, 225
Determinability, 56
Digital signal processor (DSP), 7
Direct current (DC), 96
Direct methanol fuel cell (DMFC), 100

E
Electro active polymers (EAP), 90
Enigma machine, 63, 64, 65, 68, 69, 70

F
Frequency capacity, 121
Friction Stir Processing (FSP), 224, 225, 233, 243
Friction surfacing (FS), 224
Friction Surfacing (FS), 224, 243
Functional graded materials (FGMs), 224

G
Gear member, 21, 22, 23, 24, 28, 29, 30, 33, 34, 37, 45, 46, 47, 48, 49
Graphical user interface (GUI), 7

H
Headstock's, 121
High speed milling (HSM), 5
Homogenisation, 224, 225
Human articulation, 59
Human–computer interaction, 57

I

Intelligence, surveillance, and reconnaissance (ISR), 75
Intelligent manufacturing system, 121
Internal combustion engines (ICE)., 97
Internal state, 59, 61, 64, 65, 68, 69
Isomorphous, 55, 62, 63, 70

L

Liquid propellant rockets, 169

M

Machine tool, 117, 118, 119, 120, 121, 128, 165, 166, 167
Machining centers (MCs), 5
Manufacturing process, 225
Maximum takeoff weight (MTOW), 74
Metal matrix composite manufacturing (MMCs), 224
Micro- and nanoelectromechanical systems (MNEMS), 109
Microelectromechanical systems (MEMS),, 4

N

Non-trivial behaviour, 61
Non-trivial machine (NTM), 55, 60

P

Phase locked loop (PLL),, 102
Pinion member, 21, 22, 23, 24, 30, 37, 39, 42, 43, 44, 45, 46, 49, 53
Polynomial expression, 22, 24, 28, 29, 30, 32, 33, 47, 49, 51
Proton exchange membrane (PEM), 100

R

Recrystallized, 225, 228, 229, 234, 236

S

Seal performance, 171, 196, 197, 198, 215
Space transport system, 169, 170, 203
Spindle-Bearings System (SBS), 117
Stereotypical engineer, 61
Structural analysis, 61
Surface engineering, 223
Symmetrical setup, 65

T

Tooth surface, 21, 22, 23, 24, 25, 26, 27, 28, 29, 30, 31, 32, 33, 34, 37, 39, 40, 41, 42, 43, 44, 45, 46, 47, 49, 50, 51, 52
Tooth surface form, 22, 24, 25, 30, 32, 33, 37, 39, 40, 41, 42, 49
Tooth surfaces mesh, 25
Transformation, 65, 66, 67, 68
Translation movement, 225, 229
Transmission error, 21, 22, 24, 33, 38, 39, 40, 41, 42, 49
Turbopump, 169, 170, 171, 172, 173, 174, 175, 176, 177, 178, 179, 180, 185, 187, 188, 191, 192, 193, 194, 195, 196, 199, 200, 201, 202, 203, 204, 209, 215, 217, 218

U

Unmanned air vehicles (UAVs), 74

Index

V
Vertical take-off and landing (VTOL), 84

Viscoplastic, 225, 231, 240, 242, 244